TITANIC
Those in Peril On The Sea

By

STEVE ORLANDELLA

From the Author

Stevespeak:
Three Years on Facebook (2012)

Titanic:
Those in Peril on the Sea (2013)

The Game:
A Baseball Companion (2013)

Burden of Proof:
A Vic Landell Mystery (2014)

Capital Murder:
A Vic Landell Mystery (2014)

Marathon Murders:
A Vic Landell Mystery (2015)

Dance with Death:
A Vic Landell Mystery (2015)

Midtown Mayhem:
A Vic Landell Mystery (2016)

To my Mother

Therese

The Real Historian
In the Family

White Star Line advertisement in advance of *RMS Olympic*

The author gratefully acknowledges the co-operation of

Dr. George J. Igo,
Professor emeritus, UCLA Department of Physics
- for his help on the force of the collision.

Plaintiff attorney Jay W. MacIntosh, Esq.
Defense counsel Joe Klinger, Esq.
- for their assistance in the Litigation chapter.

Harland & Wolff
- for information regarding ship construction.

Titanic Historical Society
- for its never-ending supply of on-line information.

The New York Times, April 14, 1912

Table of Contents

Foreword	9
Overture	15
The Germans and Cunard	21
The Irish Masters	45
Down to the Sea	77
April Crossing	105
Shipmates	127
Forewarned	149
Bump in the Night	169
Twenty Boats	193
The Abyss	249
Adrift	259
Deliverance	267
An Ocean Apart	279
Inquiring Minds	305
Aftermath	333
The "L" Word	355
Dénouement	363
F. A. Q.	367
Destiny	379
Bibliography	399
Index	403

> "Eternal Father, strong to save,
> Whose arm hath bound the restless wave.
> Who bidd'st the mighty ocean deep
> Its own appointed limits keep.
> Oh, hear us when we cry to Thee,
> For those in peril on the sea."
>
> William Whiting

Foreword

"Life can only be understood backwards;
But must be lived forwards."

Kierkegaard

In the fall of 1960, I was a ten-year-old, growing up in Los Angeles' San Fernando Valley. Even then I was sarcastic, opinionated, and well on my way to becoming obnoxious. The phrase most often used was, *"A little too smart for his own good."* Perhaps. Duplicit in all this were my parents, who spoiled me rotten. One of my numerous privileges was permission to stay up late on Saturday night…very late.

Toward the end of the 1950s, television in Los Angeles was in a state of flux. The Country's number three [now number two] market had seven stations, a wealth of airtime and a dearth of programming. The three network affiliates and the four independents turned to motion pictures to fill the void, so much so that one station, Channel 9, ran the same movie every night for a week. Hey, I love Jimmy Cagney, but how many times can you watch *"Yankee Doodle Dandy?"* The stations also had the nasty habit of cutting the films to pieces, the classic case being Channel 7, the ABC affiliate who filled their 3:30-5pm slots by slicing and dicing 2-hour movies down to 67 minutes. They came close to cutting Ingrid Bergman out of *"Casablanca."*

Channel 2, the CBS Affiliate, had no such problem. [They had *"Lucy;"* they had *"Jackie Gleason."*] *"The Fabulous 52"* was reserved for Saturday night at 11:30pm, and, since the only things that followed the movie were the National Anthem and a test pattern, they ran uncut. The station held the rights to a package of relatively recent films from 20th Century Fox. One Saturday afternoon my dad announced, *"Titanic is on tonight."* I had no idea who or what was *"Titanic,"* but we gathered in the family room at 11:30. For the next two hours, I sat transfixed, mesmerized by what we were seeing. If you are scoring at home, it was the 1953 version with Barbara Stanwyck, Clifton Webb and a young Robert Wagner. They had me.

In 1964, I came across a copy of *A Night to Remember,* Walter Lord's seminal work on the events of April 14-15, 1912, and the following year, saw the movie made [in England, 1958] from Lord's book. It was a film made by people who wanted to get it right. This was the game changer. The Fox movie opens with a page of text proclaiming that all the facts in the film were taken right from the United States Senate and British Board of Trade Inquiries. *"Really?"* Even then, Fox knew how to *"play fast and loose with the truth."* As good as their movie was, and it was good, it paled before the Brit's film. Fifteen hundred people did not all stand together, sing *"Nearer My God To Thee,"* and meekly sink into the North Atlantic. They fought and struggled until their last breath, trying not to freeze or drown in the unforgiving sea. Madeleine Astor wasn't an elegant matron. She was in fact, a pregnant teenager. That was it, *"Game On!"* I absorbed every book I could find, any TV program I could watch, and every newspaper on microfilm, along with help from the *Titanic Historical Society*. Add that to my natural affinity for ships, and an obsession was born. For some it's The Civil War, for others it's the Kennedy Assassination, for me it is The Royal Mail Steamship *Titanic*.

Part of the obsession stems from the fact that no event in history is so loaded with conjecture, myths, and downright lies as the wreck. Some of which are "beauties." One example: A young David Sarnoff [co-founder of RCA] became famous telling the world how he was the first to pick-up the *Titanic's* distress call in the station on the roof of *Wanamaker's* Department Store and how he remained at the key all Sunday night and well into the next day. Great story? Absolutely. Truthful story? Absolutely not. *Wanamaker's* was closed on Sunday, and even when the store was open, Sarnoff was the office manager. Three other employees of *The Marconi Company* stood the watch.

Fox reloaded and fired again in 1997. This time they tried it with a seemingly unlimited budget and an "amateur" historian calling the shots. Movie making? Unmatched. Story telling? Not so much. History? Nonexistent. There is a word for what you wind up with when you invent the leading characters. Fiction. Now, nobody loves Kate Winslet *"in flagrante delicto"* more than I do, but the truth is better. Thus, *"Jack Dawson"* and *"Rose DeWitt"* join *"Julia Sturges"* and *"Lady Marjory Bellamy"* as mythical creatures on a real ship.

And since you're making stuff up, how about a little character assassination? The 1997 film depicted First Officer William Murdoch taking but ultimately rejecting a bribe from make-believe villain *"Caledon Hockley."* Murdoch was also shown shooting two passengers dead after he presumed they intended to storm one of the remaining lifeboats. He then salutes Chief Officer Henry Wilde and commits suicide with a revolver. None of this ever happened. After the picture's director [name withheld] refused to take out the bogus scenes, studio executives flew to Murdoch's hometown to issue his relatives an apology. As for the movie, if you are looking for an accurate depiction of events - keep looking. Put another way, there was a ship called *Titanic,* and it sank. After that, you're on your own.

The Civil War is far and away the all-time champion of most books. [One of *Titanic's* passengers wrote *"The Truth about Chickamauga."*] Second? The runner-up is World War II. Third? The correct guess is the *Titanic*. So, what is my mission statement? *"What else?"* Write yet another book. Now we tell her story, once again. This time we come armed with all we knew and all we have learned in the wake of Doctor Robert Ballard's stunning discovery of the wreck in 1985. We will attempt to detail what is correct and dispel - whenever possible - at least some of what is not.

I spent my career working in television, the first seven years producing TV News. What did I learn? I learned skepticism, tinged with a bit of cynicism, and it has served me well. So, I will do your bidding. On your behalf I will be skeptical, factual, analytical and when required, cynical. There is one thing I cannot be, dispassionate. I will stipulate to a love of all ships - but her most of all. By now you may be asking yourself, *"Why so many pictures?"* I confess that too is the TV producer in me. You always try to put a face with a story, plus there is always the possibility that you can't recognize *Turbinia*.

If I am standing at all, it is on the shoulders of some truly great authors. I have read, re-read, and re-re-read their work over the years and have researched – borrowed - from them all. To the best of my ability, everything in this book is true. I believe in the concept that, if the Lord wanted us to remain silent, he wouldn't have given us [brackets]. So, on occasion you'll see a comment from yours truly. [I'll be that most irritating of shipmates, the loud, opinionated one.]

The longest section of the book concerns the area around the Boat Deck between midnight and 2:20am. If it seems long [it's real time] and overly detailed, I apologize, but to me this is the heart of the narrative. Hundreds of little dramas, played out on a sloping deck in the middle of a freezing ocean. Loved ones were torn apart, and families were destroyed. And with it came the sub-plots. Some got in lifeboats and some did not. Some were allowed in the boats and some were not. All of this begs the question: "Why?" Regardless, these are their stories, and on their behalf, I will make no apologies.

I have tried to keep the technological parts under control, and not drown my readers in facts and figures - but the brains and skill that created the *Olympic*-class liners are very much a part of this story.

Allow me just a couple of more thoughts before we proceed. There is one sentence that is common to virtually every book written about the *RMS Titanic*. *"It had been a mild winter in the Arctic."*

It had, indeed. Ice that had been forming since well before the dawn of man was now at last free. Unfettered, it could leave Greenland and move into the Labrador Current and begin its journey south toward the shipping lanes. The ice was no different than previous years, only this year there would be more than usual - much more. There were small pieces of ice, what sailors called *"growlers."* There were large sections, known as *"sheet ice,"* and larger still, *"pack ice."* In between were hundreds of what every seaman feared most, what the Norsemen referred to as *"mountains of ice."* Icebergs.

If you're familiar with the advertising business, you probably know about the concepts of *"marketing research"* and *"brand recognition."* Countless studies have been commissioned to find out what people can identify and what they like. The results are often quite surprising. For example, inquiries have determined that far more people [around the world] can recognize the *"Cavallino Rampante"* [in English, *"The Prancing Horse"* aka the *"Ferrari"* logo] than can recognize *"Shell"* or *"Coca-Cola."* Then there is my favorite. For decades focus groups, when asked to identify the most famous ship in the world, gave the traditional answer, *Noah's Ark*. No more. The runaway number one is now the *Titanic*. That's *"brand recognition."*

There is no way to tell the whole story in this little book, yet we will do our best. Call me crazy [you wouldn't be the first] and maybe a little arrogant [see previous], but I feel it's my duty to help set the record straight for fifteen hundred souls who went to a cold, watery grave that night. Time to depart. *"All ashore that's goin' ashore!"*

Overture

"Unless you try to do something beyond what you have already mastered, you will never grow."

Ralph Waldo Emerson

Queen Victoria & Prince Albert [circa 1860]

"We are living at a period of most wonderful transition which tends rapidly to accomplish that great end to which indeed all history points, the realization of the unity of mankind."

Prince Albert

It was called "The Golden Age." Under the rule of Queen Victoria and her consort Prince Albert, England was at the forefront of the Industrial Revolution. In spite of its lack of natural resources, [except for coal] the combination of political peace at home and new technologies, would transform what a sneering Napoléon dubbed, *"that damnable little nation of shopkeepers"* into an economic powerhouse. Aggressive colonization would bring markets around the world under its influence, a Commonwealth of Nations, protected by [Bonaparte again] *"their thrice-damned navy."* And with it came a kind of "peace." History would call it the *Pax Britannica,* the belief that all the world's problems could be solved by commerce. It was Capitalism on a grand scale, with capable Brits calling the shots, and fortunes made overnight. The prevailing sentiment was all positive, all optimistic, after all *"The sun never set on the British Empire."* Their economic dominance would be unchallenged until a former colony from across the Atlantic came crashing onto the scene.

Following the Civil War and the opening of the west, the United States sat on a treasure trove of natural resources. The Industrial Revolution took hold in America as well. They were rivals to be sure, however, there was a peaceful co-existence between the two countries, and both economies thrived. The British had written the book, and the Americans had studied hard. *Manifest Destiny* had driven the U. S. westward for a half century; now those same principles would propel her in the opposite direction, toward the sea.

> *"Whoever commands the sea, commands the trade; whosoever commands the trade of the world commands the riches of the world, and consequently the world itself."*
>
> *Sir Walter Raleigh*

The possibilities seemed endless, but what about the price tag? The emergence of huge corporations, welcomed by some, was most troublesome to others. Post-Civil War "friendly legislation" and the lack of an income tax laid the groundwork for what lay ahead. Under J. P. Morgan, U.S. Steel owned land equal to nearly half of New England. Albert's dream of peace and prosperity was becoming bastardized into mere acquisition. Now wealth itself was the goal, a religion of sorts, its own *Gospel of Mammon.*

The "Golden Age" had become, as Mark Twain famously labeled it, *"The Gilded Age."* The period, while glittering on the surface was corrupt beneath. The late nineteenth century would be remembered primarily as the period of rapacious Robber Barons, unscrupulous speculators, and corporate buccaneers. All of this played out against a backdrop of shady business practices and scandal-plagued politics. It was as if greed and guile had been raised to an art form.

And with it came the vulgarity of displaying your wealth. Not only did you need a grandiose mansion in New York or Philadelphia, or a glittering townhouse on Boston's Beacon Hill, but also a summer *"cottage"* in toney Newport, Rhode Island. Above all came arrogance. The thinly veiled comparisons to Greek Tragedy were on display for all to see. Here were the modern day epic heroes, the demigods complete with *hubris,* worshipping at the temple of progress. They paid homage to the notion that man could do anything, and with their machines, had finally overcome every obstacle. Even nature.

On the morning of October 21, 1805, twenty miles off a small cape along the south coast of Spain an English Admiral, having pursued his enemy's forces across the Atlantic and back, led twenty-seven *"ships of the line"* against the combined fleets of France and Spain.

Outnumbered and outgunned the Royal Navy would do what they had always done - seize the initiative. Three hundred years of naval tradition had created an élite officer corps. Now these men, the heirs of Francis Drake, John Hawkins and Martin Frobisher formed their ships in *"line ahead"* behind *HMS Victory*. The British Captains - already immortalized by their leader as his *"band of brothers,"* guided by an ingenious battle plan and with the wind at their backs, broke the defender's line and delivered an overwhelming assault. Of the thirty-three *men-of-war* in the Franco-Spanish order of battle that day, eighteen would be captured and the rest scattered or sunk.

It would be the capstone in the age of *"fighting sail,"* a crowning moment for *"wooden ships and iron men."* It meant nothing less than the salvation of England from invasion by Napoléon, and a guarantee of Britannia's domination of the seas for the next century. As always, war comes at a price and for all this the *"little nation"* would pay dearly - the loss of history's greatest Fleet Commander. The Cape was Trafalgar. The Admiral was Nelson.

This is what followed.

Cunard advertisement poster, Early 20th Century

Chapter 1
The Germans and Cunard

*"Rule, Britannia!
Britannia, rules the waves!"*

James Thomson

The events of April 1912 had their genesis two centuries earlier. The most competitive and most lucrative sea-lane in the world is the North Atlantic Ocean. As the British colonies in the Western Hemisphere flourished, so did the traffic from Liverpool and Plymouth to New York and Boston. The revolution that made the *"thirteen sisters"* a nation did nothing to slow the traffic. Cargo and passengers moved, as always, in sailing ships known as *"packets."* They moved through some of the most tempestuous waters on the high seas, across an ocean replete with fog and ice. There was no such thing as a regular service. Ships of that era waited for a full load and a destination. What today would be known as *"tramp steamers."*

A shipping revolution was coming, and it would not be centered on either freight or people, but rather something else - mail. The British Government was taking bids for a contract to deliver letters and packages across what the English called the *"Western Ocean."* In 1839, the first transatlantic mail steamship contract was awarded to a Canadian entrepreneur. Within a year, he formed the British and North American Royal Mail Steam-Packet Company.

A mail delivery service was not enough for him. He wanted one thing more, the first scheduled service on the North Atlantic. Once a week, on the same day and at the same time, a ship would leave for America, and another would sail home. He calculated that it would take four ships to maintain this schedule. He then went one step farther; he equipped his ships with sails and the latest development - a coal-fired steam engine. In 1840, the company's first steamship, the *Britannia* sailed from Liverpool to Halifax and on to Boston. [Massachusetts was a full day closer to England than New York]. On board were the owner and 63 passengers, marking the inauguration of his passenger and cargo service from the Old World to the New.

Samuel Cunard, Founder, The Cunard Line

RMS Britannia, **The Cunard Line**

From day one, his ships would establish a reputation for absolute reliability, with Captains sailing on orders that demanded one thing:

> *"Your ship is loaded, take her, speed is nothing, follow your own road, deliver her safe, bring her back safe - safety is all that is required."*

It would be seventy-five years before anyone came to a violent death on one of his ships, and then not the result of ice, but rather a German torpedo. The *Britannia* was followed by the *Arcadia*, *Caledonia*, and *Columbia*. The *"Atlantic Ferry"* was born. He would die a Baronet on April 28, 1865, leaving in his wake, nine children and the most celebrated name on the high seas. Cunard.

"Uneasy lies the head that wears the crown." The New York & Liverpool United States Mail Steamship Company, better known as the Collins Line, loomed as a serious challenger. From the time of *"The Boston Tea Party"* through the War of 1812, Americans had sought ways to tweak the lion's nose on the seas whenever possible. In 1849, the U. S. Postmaster General Office invited companies to submit bids for a 10-year federal government-subsidized mail contract between New York and Liverpool. The service would be in direct competition with Cunard, which had [finally] been on the New York run for over a year. An American, Edward Collins, submitted his ambitious plan to operate a weekly service on the route with five [later changed to four] ships superior to those of Cunard in [almost] every way. Collins won the battle, receiving his subsidy, but ultimately lost the war. His ships, *Atlantic, Arctic, Baltic* and *Pacific* were faster than Cunard's, but that came at a price. Collins' Liners used twice the amount of coal as the Brits, and the excessive engine power resulted in numerous failures. The real problem was safety.

SS *Arctic*, Collins Line

SS *Pacific*, Collins Line

In September of 1854, the *Arctic* collided with the French steamer *Vesta* in the fog off Cape Race. She had no watertight bulkheads and sank with a loss of 233 passengers. Two years later, the *Pacific* disappeared without a trace while on a voyage from Liverpool. It was believed that she hit an iceberg and 240 people perished. Congress then acted to turn off the spigot, unwilling to bankroll anymore of Collins' *"death rides."* A century would pass before an American line competed successfully on the North Atlantic, while the Liverpool firm would face a more serious threat. The last challenge had come from across the ocean; this new one would come from across the harbor.

In January 1868, Thomas Ismay, a director of the National Line, purchased the house flag, trade name, and goodwill of the bankrupt shipping company for £1,000. With these meager assets, he planned to operate large liners on the Atlantic run. On January 18th, Ismay was approached by Gustav Schwabe, a prominent Liverpool merchant, and his nephew, shipbuilder Gustav Wilhelm Wolff. Wolff was the part owner of a shipyard in Belfast, Northern Ireland. Schwabe pitched his proposal. He would finance the building of Ismay's ships on the condition that his nephew's company, Harland & Wolff, would build all of the liners. Ismay agreed, and a relationship with the shipbuilders was established. William Imrie came on as a partner and together they formed the Oceanic Steam Navigation Company. The house flag was a red pennant with a single white star, heralding the new White Star Line. A century before it was *en vogue*, Ismay had learned the value of brand recognition. By design, the names of Cunard's ships ended in *"ia."* [*Mauretania* or *Lusitania*.] Not to be outdone, White Star's ships all ended in *"ic."* [*Majestic* or *Oceanic*.] Since the *Britannia*, the funnels on all Cunard Liners had been painted vermillion and black [in fact, they still are].

Thomas Ismay, Founder, White Star Line

***RMS Oceanic*, White Star Line**

White Star opted for a more elegant buff gold and black. Thomas Ismay had thrown down the gauntlet.

The Line, now holding its own Royal Mail contract, hit the ground running with no less than six new ships. Harland & Wolff launched the *Oceanic, Atlantic, Baltic* and *Republic,* and then followed with the *Celtic* and *Adriatic*. In the latter part of the nineteenth century, White Star operated many famous ships, *Britannic, Teutonic,* and *Majestic*. At one time or another several held the *Blue Ribband*, the mythical prize awarded to the ship making the fastest crossing.

During this period, Marine Engineers learned a valuable lesson: Speed is expensive. Velocity increased in logarithmic proportion to engine power and fuel consumption. Simply put, above twenty-two knots [25 mph] reciprocating steam engines required very high power and an enormous amount of coal. Plus, every knot of speed above twenty would cost a ship owner as much as the first twenty had in research and development, design, testing and construction. It was at this point White Star rewrote the game plan. No more chasing after the *Blue Ribband* for them. From now on, the Line would opt for comfort, reliability, and economy of operation over speed.

In the second half of the nineteenth and the early twentieth centuries, upwards of thirty million people migrated from Europe to the United States and Canada. White Star was one of the first shipping lines to operate passenger ships with inexpensive accommodations for Third Class passengers, in addition to their usual spaces for more affluent First and Second Class travelers. The Line would lead the way in offering *"little luxuries"* to their newest passengers, and their ships now had room for up to one thousand in their lowest class of service. Thomas Ismay had proven to be a very worthy adversary.

Perhaps a little too worthy, as this encroachment on their business didn't sit well with the front office at Cunard. The rivalry between the Liverpudlians was reaching new heights with even more competitors arriving on the scene. The *Compagnie Générale Transatlantique,* [translation *"The French Line"*] was building liners [*"Mon Dieu!"*].

One of the visitors to Britain's Naval Review of 1889 was the new German Emperor, Wilhelm II. Since he was a grandson of Queen Victoria, he was welcomed with open arms. While at Spithead, he saw all the Royal Navy's capital ships and was duly impressed. Something else caught his eye, the new White Star Liner *Teutonic*. The Kaiser was taken by her size, her power and her luxury. Wilhelm had seen quite enough, *"Die Würfel sind gefallen"* [*"The die is cast"*]. Germany would enter the contest and claim a share of the profit and prestige to be had on the North Atlantic. The Kaiser watched and learned. His schooling taught him that there were two great pillars on which the British Empire was built, the Royal Navy and the Merchant Marine [three, if you counted *Lloyd's of London,* who underwrote the whole thing]. If Imperial Germany meant to challenge the *Pax Britannica,* it would be done on the high seas. [How does *Pax Germanica* sound to you?]

In just eight years, the Germans were ready - very ready. With the help of big subsidies from the government, the *Kaiser Wilhelm der Grosse* of the Norddeutscher Lloyd Line entered service and captured the *Blue Ribband*. Not to be out done, Germany's other line, the Hamburg-Amerika Line [HAPAG for short], put the *Deutschland* into service. She promptly took the *Blue Ribband* back, but lost it again when the Bremen firm unveiled the still quicker *Kaiser Wilhelm II.* [Good idea to name the ship after the guy who's writing the checks.] The *"Kaiser"* featured the largest reciprocating engines in the world.

SS Deutschland, Hamburg-Amerika Line

Kaiser Wilhelm II, North German Lloyd Line

German industrial power had been put to good use. The quest for the North Atlantic speed record was now a Hanseatic volleyball game. For the proud Kaiser and his new fleet it was *"Volle kraft voraus!"* [*"Full speed ahead!"*] For the English, it was *bête noire*.

At the turn of the century, the battle for shipping supremacy raged between the express liners of England and Germany. The traditional dominance of the sea-lanes by Great Britain had been challenged by an upstart sea power from across the North Sea. The four fastest ships on the *Atlantic Ferry* had one thing in common - they were German. The *Blue Ribband* was now securely in their hands. British prestige was under attack. Imperial Germany was building a High Seas Fleet to challenge the Royal Navy. To a nation that survives only because of its supremacy at sea, the presence of a burgeoning industrial power with aspirations on the water was cause for concern.

Simply put, to the British, control of the oceans meant life itself. In England, if you won a great victory on land you rated a train station [Waterloo]. Win one on the sea? That got you a square [Trafalgar].

There was still more trouble looming on the portside. Enter the United States in the form of Financier Junius Pierpont Morgan. Morgan, a *"Gilded Age Buccaneer"* if ever there was one. He had conquered America's railroads and was now interested in acquiring the Atlantic. His International Mercantile Marine Company [IMM] was a holding company that controlled subsidiary corporations. His plan was to dominate transatlantic shipping through contractual arrangements between the lines and the roads. Thus creating on the high seas what he had already fashioned on the rails - a monopoly. Why build commerce if you can buy commerce? At 219 Madison Avenue, *The House of Morgan* was opening its checkbook.

J. P. Morgan, Owner, International Mercantile Marine

Sir Charles Parsons, Inventor of the Marine Turbine

By 1903, the IMM had acquired all of the top firms with the exception of Cunard and the French Line. Most disquieting to the British was their takeover of The White Star Line. The Line, along with Cunard, were more than just the nation's leading shipping companies. In time of war, their ships would be pressed into service as armed merchant cruisers. Now, the Admiralty would need the permission of an American conglomerate to put their vessels *"in harm's way."*

The double threat to British pride and British security could not be tolerated. It was particularly galling to Cunard to see its place of honor lost to a bunch of *"Johan-come-latelies"* from Bremen and Hamburg. Into the breech stepped Lord Inverclyde, Chairman of Cunard. Seeing an opening, he approached the Balfour Government [the Earl of Balfour, Prime Minister from 1902 to 1905] with a request for a loan [translation *"subsidy"*]. The money would build two express liners, faster than anything on the water, and naturally available to His Majesty's Navy in time of war. In short order the funds were appropriated and the contracts given to two of England's finest yards. Scotland's John Brown & Company would build the *Lusitania,* while Swan Hunter & Wigham Richardson would construct the *Mauretania* at Wallsend by the River Tyne.

Cunard wanted luxury ships. The Admiralty wanted dependable ships, safe ships, and above all, fast ships. Safety would be guaranteed by the layout of her engine spaces. Coalbunkers would be aligned along the sides of their hulls to protect the engines from enemy [pronounced *"German"*] gunfire in wartime. If holes where punched in a ship's side, she would still float. Further protection would come from a "double bottom," with the outer hull absorbing the damage and protecting the inner hull, while transverse bulkheads would control flooding. The speed would come from somewhere else.

Steam Yacht *Turbinia*

***HMS Dreadnought*, first turbine powered battleship**

The design specifications called for a maximum speed of no less than 24 knots, a huge leap forward in performance. Reciprocating engines would not be the answer. To reach that goal would require the adoption of an all-new, and very different form of motive power. This brings us to a man named Charles Algernon Parsons.

Parsons was in the process of forever changing ship travel with his invention, the Marine steam turbine. The turbine was a smoother and far more efficient means of generating propulsive power than traditional expansion engines. His genius for engineering was only matched by his gift for publicity. Seeking to land a lucrative contract from the Admiralty, he built a test bed one hundred feet long, nine feet wide, and christened her *Turbinia*. She stormed her way into the history books on the morning of June 26, 1897, when Parson's brainchild gate crashed the Naval Review at Spithead, for Queen Victoria's Diamond Jubilee. On hand were the Prince of Wales, Lords of the Admiralty, and foreign dignitaries including, of course, "*das Kaiser*." On Parson's command, *Turbinia*, far faster than any other ship of the time, entered the anchorage. She raced between the two lines of battleships, steaming up and down in front of the crowd and princes with impunity. Royal Navy destroyers were dispatched to intercept her. This was a hopeless endeavor when you consider their best speed was 27 knots, while the steam yacht could exceed 34 knots. The next day, the very proper *Times of London* reported:

> "Her lawlessness may be excused by the novelty and importance of the invention that she embodies."

The not so proper *Daily Mail* got right to the point:

> "If that shrimp of a turbinet comes to anything, all these black and yellow leviathans are done for."

The Competition: Turbine rotor, *RMS Lusitania*

The Competition: Double bottom, *RMS Mauretania*

The Competition: Launch of *RMS Lusitania* in Scotland

The Competition: Fitting out, *RMS Mauretania*

The Competition: *RMS Lusitania*, Pier 54, New York

The Competition: *RMS Mauretania* arrives in New York

"The Mail" had got it right. A star was born. The Admiralty was sold on the idea of steam turbines for its ships, leading to 1906 and the world's first turbine-powered battleship, *HMS Dreadnought* [thereby making every other capital ship in the world obsolete]. The Line had found its power source. Parson's better mouse trap was the answer to the question of speed. Now, the normally conservative Cunard took a chance. They would roll the dice on Parsons and his amazing engines to retrieve the *Blue Ribband,* fully aware that turbines of the required size had never before been placed in a ship.

RMS Lusitania was launched into the River Clyde in June 1906, and entered service in September 1907. Known to her crew affectionately as *"The Lucy,"* she was a hit right from the start. As luxurious and far faster [24.5 knots] than any other liner on the New York run, she was hugely popular. Cunard's faith in [soon to be *"Sir Charles"*] Parsons was justified. The arrival of *RMS Mauretania* would be delayed until November of 1907. The engineers at Swan Hunter were not satisfied. They believed an improved hull design could yield even more speed. The yard made their own scale model and tank tested relentlessly, until the staff was convinced they had the perfect design. It was time well spent. *"The Maurey"* could run at a previously unheard of 25.75 knots - if she wasn't perfect, she was darn close. Other shipbuilders were staggered and even Rudyard Kipling was impressed.

> *"You can start this very evening if you choose*
> *And take the Western Ocean in the stride*
> *O seventy thousand horses and some screws!*
> *The boat-express is waiting your command*
> *You will find the Mauretania at the quay,*
> *Till her captain turns the lever 'neath his hand,*
> *And the monstrous nine-decked city goes to sea."*

Cunard advertisement poster in advance of *RMS Aquitania*

The *Blue Ribband* was hers. She would hold it for 22 of the next 28 years. The German racers had been hopelessly outclassed. Cunard had [just as Nelson a century before] *"swept the seas for England!"*

Cunard was now holding two Queens and still had an ace up its sleeve. The new superstars supplanted two older liners on the Atlantic run. It still required four ships to fulfill the weekly schedule. That would change as well. The Line, this time using its own money would build a third, larger turbine-driven ship. The keel of *RMS Aquitania* was laid in 1910. Once again, it would be John Brown and Company building her on the Clyde [they of the *Lusitania*]. Such was the speed of these ships that, when she entered service in 1914, Samuel Cunard's magic number would be reduced from four to three.

In August 1851, an American-built schooner sailed across the Atlantic to compete in a race around the Isle of Wight. The three-master was pitted against a flotilla of England's fastest yachts. With the Queen and the Royal Yacht Squadron in attendance, their guests from New York crushed the "home team." The *America* took the prize back to the United States [thus was born *"The America's Cup"*]. A definitely *"not amused"* Victoria then asked, *"Who finished second?"* An aide replied, *"Your Majesty, there is no second."* Now Cunard was first and there was no second. White Star could no longer just stand and watch. It was time to stand and deliver. They would have to respond.

They did. [Morgan's money spent just as well as Admiralty money]. On April 30, 1907, a Daimler-Benz town car elegant in [Roi-de-Belge] custom coachwork stopped in front of No. 27 Chelsea Street in the Belgravia section of London. A footman ushered Mr. and Mrs. Joseph Bruce Ismay into the back seat and their chauffeur drove the couple to No. 24 Belgrave Square, home of Lord and Lady Pirrie.

J. Bruce Ismay, Managing Director, White Star Line

Lord Pirrie, Chairman, Harland & Wolff

The Ismay's were to be the dinner guests of the Pirries. Joseph, the son of White Star's founder, was the Managing Director. William James Pirrie [1st Viscount Pirrie] was the senior partner and Chairman of the Board of Harland & Wolff. The Belfast firm still constructed all of White Star's vessels. Their relationship had remained unchanged since the days of Ismay's father. The Queen's Island yard would not build ships for any of White Star's competitors [translation *"Cunard"*]. Everything was done on a cost plus basis. Harland & Wolff would design and build the vessels for a price, and then a fixed percentage [5%] would be added as the company's profit.

Their Merseyside rivals were about to put into service the *Lusitania* and the *Mauretania,* and word was out about their capabilities. Over dinner with their ladies the men discussed Cunard's *"pas de deux"* and how to counter the competition [with no doubt, some less than flattering remarks about the now ex-prime minister, Lord Balfour]. In Pirrie's den, accompanied by Napoleon brandy and Cuban cigars the moguls made their move, drawing up plans for the construction of three ships, each half again as large as the opposition's new liners. Ships so large that both men knew as they spoke, there did not exist a berth, a drydock, or even a pier that could accommodate them.

Ismay had no intention of challenging the *Mauretania's* speed record. Like his father, the *Blue Ribband* meant nothing to him. What his new ships lacked in speed, they would make up for in space, style and luxury. He envisioned luxury on a scale never before seen on a British ship, and comfort enough so that the few extra hours of a crossing would pass unnoticed. The *Lusitania* had cost the Admiralty [and Cunard] $4,000,000. Morgan and the IMM would have to "pony up" $7,500,000 for each of the White Star ships.

Ismay was hell bent on putting the *"luxury"* in *luxury liner,* and by evening's end, Pirrie had drawn a preliminary sketch featuring a long graceful hull, near vertical stem, two masts, and four funnels. In time, the silhouette would become the stuff of legend, as would the names, names commensurate with the liners sheer size and power. Did the irony of the situation occur to these men? That as a result of their rival's successes, they had [via Morgan's millions] been given the opportunity of a lifetime? Their own chance to *"sweep the seas."*

Their benefactors [Morgan and International Mercantile] signed off on the plan. On 29 July 1908, Harland & Wolff presented the drawings to Ismay and other White Star Line executives. He approved their designs and signed three letters of agreement two days later, authorizing the start of construction of a troika of triple-screw steam ships each with a displacement in excess of 45,00 tons. Word of the order was immediately flashed to White Star's headquarters in Liverpool and then to Alexander Carlisle [Managing Director and head of the draughting department] at the shipyard on Queen's Island in Belfast. The lead ship would be the *Olympic,* thereby making both subsequent vessels *Olympic*-class. Lastly would come the *Gigantic,* she would follow the *Titanic.* The triplets would be built to do battle with the *Mauretania, Lusitania,* and *Aquitania.*

On the evening of April 10, 1910 - and precisely on schedule - a reoccurring celestial phenomenon known to all astronomers as *"1P/Halley"* appeared in the heavens above the Harland & Wolff shipyard. *"Halley's Comet,"* brilliant in the night sky, was believed by many to be a spectacular harbinger of good luck. To others, it was a dreaded omen of impending disaster.

Chapter 2
The Irish Masters

*"Moderation is a fatal thing.
Nothing succeeds like excess."*

Oscar Wilde

So, whose signature was on the blueprints? The popular belief was Thomas Andrews. Andrews, however, whose pension for detail was all over the liners, did not become Managing Director until after the departure of Alexander Carlisle in late 1907. Carlisle was responsible for much of the design, including the watertight bulkheads and the lifeboats. When questioned at The Board of Trade Inquiry, Carlisle said that Lord Pirrie was the principal designer, and the details were left to him. The truth is that no one man can claim to be the designer of the *Titanic*, but a family can. Carlisle was Pirrie's brother-in-law and Andrews was Pirrie's nephew. There was no cry about nepotism; however, all three were very competent naval architects. Work began immediately in Belfast, with Carlisle and the draughting department clearing the decks for the biggest order the yard had ever received. Andrews' deputy, Edward Wilding was added to the design team and made responsible for calculating the ship's design, stability and trim.

Three slipways were merged into two to accommodate the width of the new liners. A gantry was built that covered the entire work. The Thompson Graving Dock, the largest drydock in the world was ready by the spring of 1911. The detritus of previous work was removed and blocks were laid on the cradle. Lastly, a giant floating crane was moved into position. [Ironically, the crane was built in Germany.] Harland & Wolff had assembled enough muscle and skill to build the first two in tandem. Shipbuilders in the truest sense of the word, virtually all of the vessels would be built on Queen's Island with very little *"farmed out."* The result was a uniformity of construction and a level of quality that few yards could match. If shipbuilding was an art [and it was], then here were the *"Irish Masters."* The *Olympic's* keel was laid in December 1908. She was scheduled to enter service in 1911, followed a year later by *Titanic*. *Gigantic* would arrive in 1914.

The Arrol Gantry, Harland & Wolff

The Drafting Room, Harland & Wolff

The *Olympic* would be built on slipway No. 2, with the hull number 400. To her right was hull number 401, the *Titanic*. Next door on slipway No. 1 the yard was building [also for White Star] the tender *Nomadic*, part of Ismay's *"little surprise"* for the people of Liverpool.

The decision not to compete with the Cunard "racers" for speed honors allowed White Star to opt for a traditional and cheaper propulsion system. *Lusitania* and *Mauretania* each had 4 propellers; the Lines new ships would have three. A total of twenty-nine boilers would provide steam for two triple expansion engines driving the wing propellers, with the center prop taking power from a Parsons low-pressure turbine [Sir Charles' fan club now included White Star]. There were, however, two differences. The *Mauretania's* propellers could be reversed; the *Olympic*-class center prop could not. In addition, a center mounted propeller takes space normally allotted to the rudder. A smaller rudder makes for a slower turning ship.

The sisters would be built in the traditional way, by the hands of the intensely proud and skilled Irish craftsmen of Harland & Wolff. Unproven technologies need not apply. First came the keel, the spine of the ship. From there protruded the "frames" [uprights] forming the skeleton. Steel plates, some more than an inch thick would become the shell plating. They began as white-hot lengths of steel, which would be wrought [some would say *"tortured"*] into shape. The plates would be attached to the hull precisely and then came the rivets. Three million bolts of hot iron "cooked" in furnaces. As they cooled they would expand, and draw the plates together.

True to the names on the blueprints, the two ships that grew under the gantry were astonishing. They were huge and yet, they were beautiful, powerful but elegant, enormous without being grotesque.

RMS Titanic & RMS Olympic on the stocks

The Irish Masters, Harland & Wolff

Reciprocating engine, *RMS Titanic*, Harland & Wolff

From Harland's grimy sprawl would come two marvels, part floating palace and part racing yacht. Sailors always refer to their ships in the feminine with the operative personal pronouns being *"her"* or *"she."* [Except, of course, the Germans, who saw their ships as *"männlich."*] That said, there was no other way to describe the two *"lookers"* rising above Queen's Island, twin goddesses wrought from iron and steel.

Implicit in the design were four smoke stacks. Why so many? Four funnels had become *de rigueur* among the first class liners. The German greyhounds had four, grouped in pairs. The *"Maurey"* and the *"Lucy"* had four, grouped forward towards the bow. The *Olympic*-class liners required only three, but a fourth, a dummy, was installed. Why? This was done for a simple reason, to answer the demands of the sea-going public. Four funnels, especially among the immigrant class, were seen as signs of strength, power, and safety. Many is the time a steerage passenger would cancel his booking, after discovering that his ship had three, or two, or [God forbid], one stack. [To White Star: *"vox populi, vox dei."*]

As work progressed on the two ships, a myriad of decisions had to be made, running the gamut from flatware to lifesaving. In their quest to operate the most comfortable ships afloat, no expense was spared. White Star's *largesse* could be seen most everywhere, except perhaps on the Boat Deck. In the later part of the 19th century and the dawning of the 20th ship size had grown almost exponentially. The laws governing lifesaving equipment were now hopelessly outdated, most notably, the ones concerning lifeboats. The largest category of ships, which of course, would include the *Olympic* and *Titanic,* required sixteen lifeboats be provided. If every boat was filled to capacity, they could accommodate 990 passengers and crew.

Lifeboats with Welin davits, Boat Deck, *RMS Titanic*

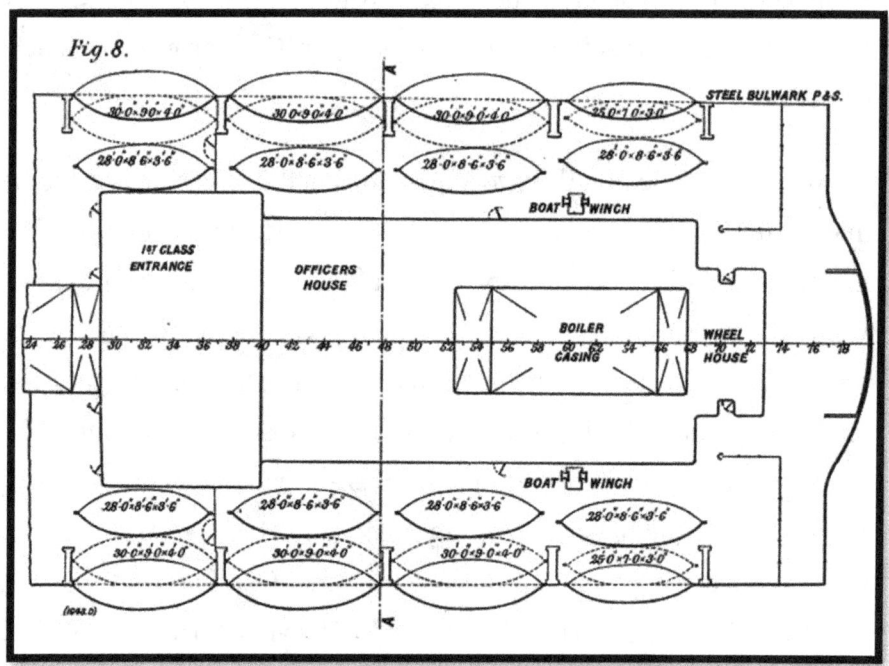

Carlisle plan for additional lifeboats, *Olympic*-class liners

That number was far less than half the total capacity of the new class of ships. Carlisle had made provision for more lifeboats. He had ordered the ships equipped with Axel Welin's new quadrant davits to facilitate up to four rows of boats. Furthermore, there was space along the Boat Deck railing for more lifeboat stations. Both Carlisle, and then later Andrews, urged the Line to add adequate boats for all. Here is a classic example of customer comfort vs. customer safety.

The Boat Deck was also First and Second Class promenade space. Someone [pronounced *"Ismay"*] at White Star decided that the Line's most important passengers should not be deprived of this area. So, for the moment, the *Olympic*-class steamers would put to sea with the minimum number of boats required by the Board of Trade. Andrews was ultimately allowed to add four smaller Englehardt collapsible boats, [forward of the promenade] bringing the total to twenty. The Line's claim that they had provided more boats than the law required would be oft repeated after April 15th, and every time it would fall on millions of deaf ears. A ship with a capacity of well over 2,500 would cross the Atlantic Ocean with enough lifeboats for 1,150.

Perhaps the Line's *laissez-faire* attitude about lifeboats stemmed from their confidence in the structure of the ship itself. No fewer than 15 transverse bulkheads divided the ship into 16 watertight compartments. *Titanic* could float with any two compartments breached, thus rendering her immune, in theory, from a broadside collision. She could float with the first four compartments open to the sea, thus rendering her immune, in theory, from a head-on collision. A White Star Line publicity brochure produced in 1910 for the twin ships *Olympic* and *Titanic* states, *"these two wonderful vessels are designed to be unsinkable."* [The game was afoot.]

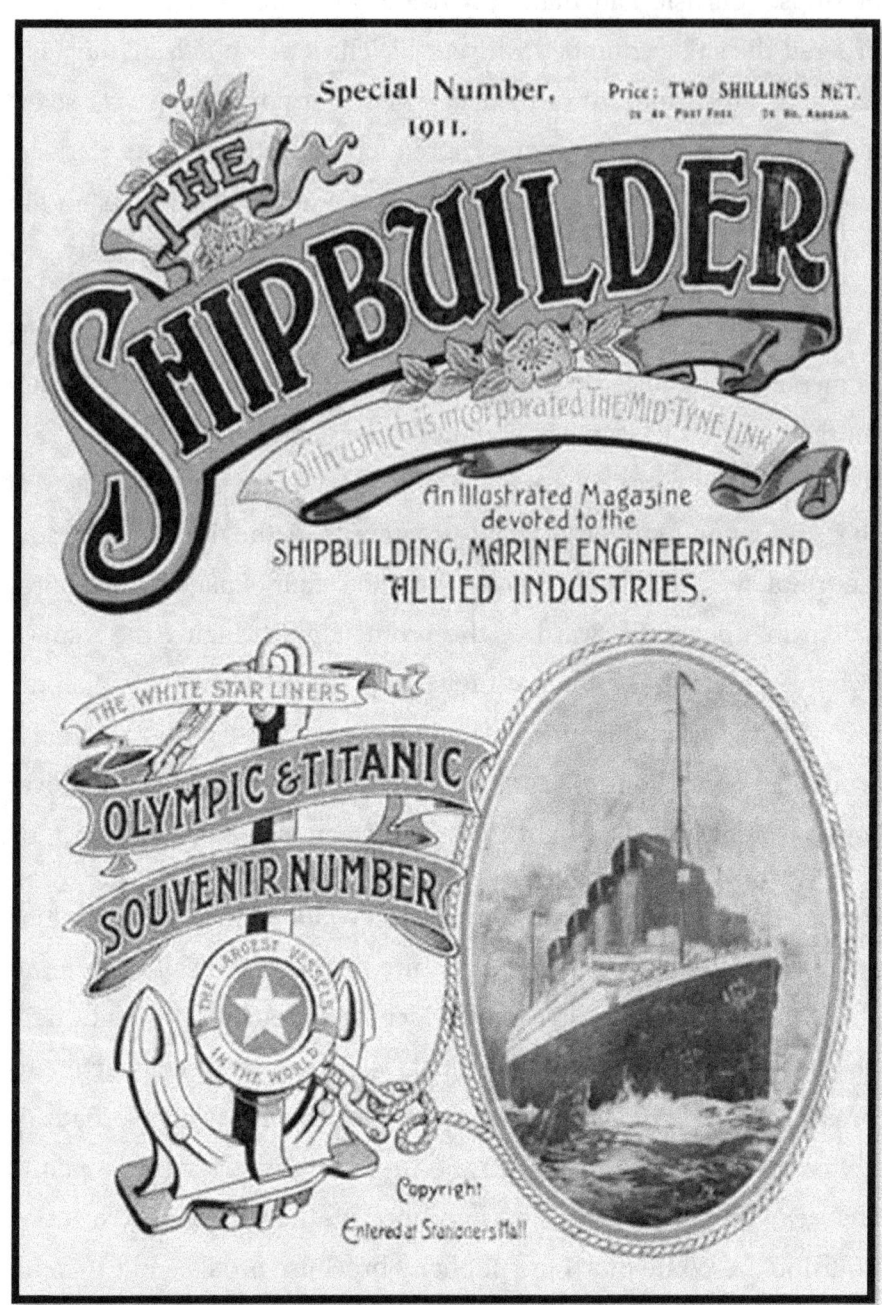

"Shipbuilder" Special Issue on *Olympic*-class steamers

In 1911, *Shipbuilder* Magazine published an entire special edition devoted to the *Titanic* and *Olympic*. One article described the construction of the ships and concluded that they were *"practically unsinkable."* The Irish News and Belfast Morning News contained a report on the launching of *Titanic's* hull. The article described the system of watertight compartments and electronic watertight doors and concluded that *Titanic* was *"practically unsinkable."* For their part, the management of Harland & Wolff insisted she was never advertised as an unsinkable ship. They claimed that the *"unsinkable"* myth was the result of people's interpretations of articles in the newspapers and magazines. Whatever the reason, the claim stuck. On that fateful night, a *Titanic* crewman tried to calm a nervous passenger by saying, *"God himself could not sink this ship!"* There was, of course, an eventuality that neither *Shipbuilder* nor the newspapers had ever considered. Their guarantees would be of small comfort to thousands of people in the days and weeks to come.

As a result of their differences of opinion about the lifeboats, Carlisle left Harland & Wolff to design *Dreadnoughts* for the Royal Navy. [The Admiralty had no Ismay.] The task of bringing the two giants to life fell to Andrews. Here indeed was the man for the job. He had worked his way up from apprentice to become the Managing Director. Andrews, the press notwithstanding, knew there was no such thing as an unsinkable ship. The torch for more lifeboats had been passed to him, and he made no bones about his feelings.

> *"Let the Truth be known, no ship is unsinkable. The bigger the ship, the easier it is to sink her. I learned long ago that if you design how a ship will sink, you can keep her afloat. I proposed all the watertight compartments and the doublehull to slow ships from sinking, so you get everyone*

> off. There's time for help to arrive, and the ship's less likely to break apart and kill someone while she's going down."

Ismay was not impressed. First Carlisle, then Andrews and now even Lord Pirrie wanted more boats. These men were veteran naval architects, and Ismay most certainly was not. He was a businessman, [translation *"landlubber: unfamiliar with seamanship and the sea"*]. He paid for the ships, and they would be built his way. Perhaps he believed his own propaganda; maybe his ships were unsinkable. At a meeting with Andrews and Pirrie, he quashed any further discussion.

> *"Control your Irish passions, Thomas. Your uncle here tells me you proposed 64 lifeboats and he had to pull your arm to get you down to 32. Now, I will remind you just as I reminded him these are my ships. And, according to our contract, I have final say on the design. I'll not have so many little boats, as you call them, cluttering up my decks and putting fear into my passengers."*

[*"The fault, dear Bruce, is not in your ships, but in yourself."*] Andrews, having failed in his last attempt for more lifeboats, took up the *"battle of the bulkheads."* The partitions in the center portion of the ship went up as far as E-Deck. Those near the bow and the stern went only to D-Deck. Here was the *"fatal flaw"* in the design. Andrews believed the bulkheads must go higher in order to guarantee ship integrity and passenger security. In addition, a compartment is never truly watertight unless it is sealed on all six sides. The tops of the *Olympic*-class compartments were not sealed. Once again, we come to the issue of customer comfort vs. customer safety. The First Class Dining Rooms on both the *Olympic* and *Titanic* were located amidships on D-Deck - positioned as close as possible to her metacentric height, to minimize the rolling and pitching of a ship.

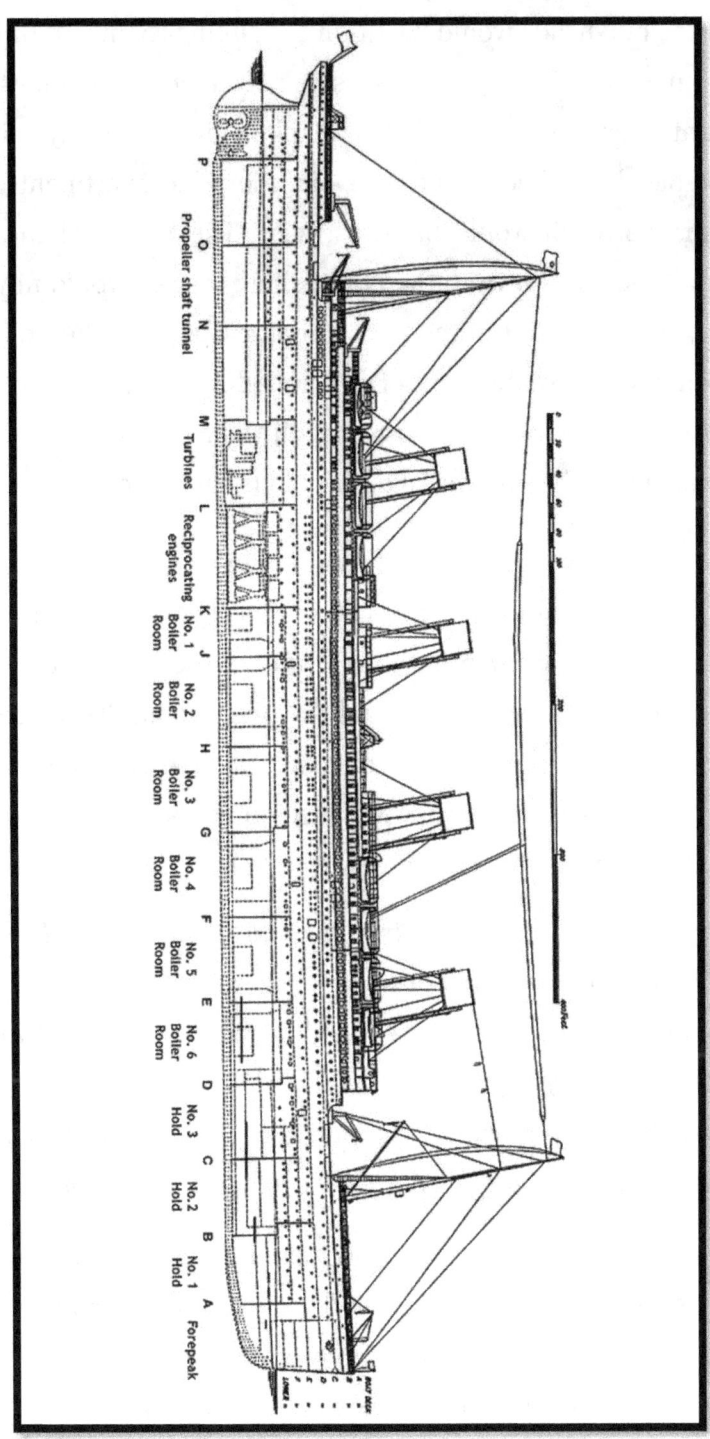

Compartments & Bulkheads (A-P) *RMS Titanic*

Raising the bulkheads would partition the Dining room and thus ruin the layout of the Salon, as well as inconvenience the Lines most important passengers. Andrews explained to Ismay, if the unthinkable happened, that five or more compartments were breeched, the result would be disastrous. The weight of the water would pull the bow down to the point where the sea would flow over the top of a bulkhead into the next compartment, and then the next, and so on until the vessel sank. The vaunted compartments would be nothing more that watertight boxes with their tops missing. White Star decided that the present layout would be sufficient.

Andrews would not give up, and then urged the Line to extend the ships *"double bottom"* past the bilges and up the side, similar to Cunard on the *Lusitania* and *Mauretania*. In the event of damage to the outer skin, the inner hull would stay watertight. These changes would drive up the price of the ships and delay their completion by months. Again White Star refused. In so doing, Ismay sealed the fate of 1500 men, women and children - one of them, Thomas Andrews.

> *"You weren't there at my first meeting with Ismay. To see the little red marks all over the blueprints. First thing I thought was: 'Now here's a man who wants me to build him a ship that's going to be sunk.' We're sending gilded eggshells out to sea. The press is calling these ships unsinkable and Ismay's leadin' the chorus. It's just not true."*

In point of fact, the reading public was less interested in lifeboats and bulkheads and more interested in luxuries and accommodations. And, as with every other facet of her design, the results were spectacular. In an era of class distinction, the comings and goings of the moneyed and titled men and women were followed with the same enthusiasm that future generations would devote to movie stars, rock stars, and great athletes. Everything must reflect a person's place in society, including the public rooms of their new favorite ocean liner. Harland & Wolff had gone all out to fulfill Ismay's mission statement. In the words of *Shipbuilder,* the *Olympic's* accommodations, *"were of unrivaled extent and magnificence and the excellent result defies improvement."* The magazine went onto say:

> *"The First Class public rooms include the dining saloon, reception room, restaurant, lounge, reading and writing room, smoking room, and the Verandah cafes and palm courts. Other novel features are the gymnasium, Squash racquet court, Turkish and electric baths and the swimming bath. Magnificent suites of rooms, and cabins of size and style sufficiently diverse to suit the likes and dislikes of any passengers are provided. There is also a barbershop, a darkroom for photographers, clothes pressing room, a special dining room for maids and valets, a lending library, a telephone system and a wireless telegraphy installation. Indeed, everything has been done in regard to the furniture and fittings to make this first class accommodation more than equal to that provided by the finest hotels on shore."*

Nothing was too good for the new liners. The interiors featured an abundance of hand-carved wooden paneling, blended nicely with crystal chandeliers, and elegant carpeting. The Line went to great lengths to insure they had the finest of everything. Even the soap *"offered a higher standard of toilet luxury and comfort at sea." "Vinolia"* was, of course, only for passengers in First Class. Those in Second and Third would have to do with something less.

Advertisement for Vinolia Otto toilet soap

Props and rudder, *RMS Titanic*, Thomson Graving Dock

Launch ticket, *RMS Titanic*, Harland & Wolff

Launch of *RMS Titanic*, Harland & Wolff, May 31, 1911

The *Olympic* was launched on October 20, 1910. The fitting-out process of installing her boilers, engines, and interiors took the better part of a year. Harland & Wolff was, far and away, the largest employer in Belfast. At any given time there could be 15,000 at work on Queen's Island. In the long history of the shipyard and White Star, it is unlikely that there was ever a more glorious day than May 31, 1911. That afternoon the yard handed over *RMS Olympic* to the Line. The main event, however, was in the morning. It seemed every man and woman in Northern Ireland was given [or took] the day off. The riverbanks were jammed and the Lough packed with boats, all to see whether the engineers could repeat what they had done the previous year, launch the largest moving object ever built by man.

There would be no off day for the yard crew at Harland & Wolff. For 24 hours they had been working, setting up the resident of slip No. 3 for her big day. Wedges had been driven in to help lift the massive hull off her cradle. The ways had been greased with tons of tallow and barrels of oil. Piles of anchor chains had been attached to her sides, to help check her momentum once the hull entered the water.

The final cosmetic touches had been made. Three-foot high gold letters on either side of her hull forward of the well deck proclaimed her as the *Titanic*. Her name, as custom, was also proudly painted across her stern just above her homeport, Liverpool. The little city on the Mersey, traditional starting point for transatlantic crossings, would never see the new liner. During construction, both Cunard and White Star announced that they were moving their first string of vessels to Southampton. This was Ismay's little surprise. They had many alleged reasons - a double tide, and better docking facilities - but the real reason was to bring the big ships closer to London and into direct competition with the Germans and the French.

The move allowed the liners to call at Cherbourg, to embark travelers from the continent. First Class passengers, always on Ismay's mind, could now take a pleasant three-hour ride on the boat train from Waterloo [rather than *schlepping* across England to the Merseyside]. Nonetheless, to superstitious sailors for whom virtually anything could be a bad omen, changing a homeport meant trouble.

White Star underplayed the moment. There was no magnum of champagne to christen her, no member of the nobility to sponsor her, no *"may God bless her and all who sail in her,"* just signal flags draped across her bow spelling out S-U-C-S-E-S-S. At the proper moment triggers fell, and, to the accompaniment of horns, whistles, and cheers, *RMS Titanic* slid into the River Lagan. It had all been a rip-roaring success, the only sour note played by *The Belfast News:*

> *"It is difficult to understand why the owners and builders named this ship Titanic. The Titans were a mythological race who came to believe they'd conquered nature, who thought they'd achieved power and learning greater than Zeus himself, to their ultimate ruin. He smote the strong and daring Titans with thunderbolts; and their final abiding place was in some limbo beneath the lowest depths of the Tartarus, a sunless abyss below Hades."*

Two weeks later on June 14, 1911, the *Olympic* left Southampton and set out on her maiden voyage. Andrews and a number of engineers and technicians were present for the round-trip passage to New York, as part of Harland & Wolff's *"Guarantee Group"* to help spot problems and areas for improvement. Few would be found. When *Olympic* tied up at the Chelsea Piers, Ismay told the assembled press, "She has done all that was expected and behaved splendidly,"

Captain Smith, Starboard Docking Bridge, *RMS Olympic*

***RMS Olympic* arrives in New York, June 28, 1911**

Shell plating damage, *RMS Olympic*

RMS Titanic & *RMS Olympic*, **Thompson Graving Dock**

April 2, 1912. *RMS Titanic* (Blue Ensign) departs Belfast

In September 1911, while outbound from Southampton and under the command of Captain Edward J. Smith, the *Olympic* collided with the British cruiser *HMS Hawke* off the Isle of Wight. Her shell plating had been pierced and two compartments flooded. The damage required a return to Queen's Island. To put her back in service, parts were taken off the *Titanic*, including a propeller shaft. In February 1912, she dropped a propeller blade, and again returned to Belfast. [The replacement did not come from the *Titanic*; their props had different pitches.] The Line authorized Harland & Wolff to again pull resources from *Titanic*, including moving her out of the drydock to make room for her sister. It would be the last time the liners were together. Her maiden voyage was delayed from March to April 1912. In spite of the extra workload, the yard completed fitting out the *Titanic* by the end of March. She left the builders on the second of April, passed all her trials, without leaving Belfast Lough and returned to the yard to drop off personnel. She then set a course for Southampton. Along with dignitaries, officers and a small portion of the crew, Andrews, with his *"Guarantee Group,"* were on-board to spot and correct any concerns and ensure a perfect first sailing.

Everything, however, was not perfect. Sometime before sailing day [no one can say exactly when] a coal fire broke out in the reserve bunker of Boiler Room No. 6. It was still smoldering when the liner reached Southampton. Coal fires were a fact of life in the era before ships became oil-fired. The Great Coal Strike of 1912 had ended, but coal was still in short supply, necessitating the "borrowing" of the now precious lumps from other White Star ships. In spite of the Lines best efforts, including cancelling sailings and transferring the coal and the passengers to *Titanic*, she would sail on April 10th with Boiler Room No. 1, containing five boilers temporarily shut down.

RMS Titanic arrives at Southampton, April 3, 1912

Lieutenant David Blair, White Star Line

There was not enough coal. In addition, there were no binoculars for the lookouts, or if there were, none could be found. Then the Line decided to shuffle the senior officers. On the *Olympic,* Captain Smith had as his Chief Officer, Lieutenant Henry Wilde. Smith decided to "borrow" him for the maiden voyage and temporarily demote Chief Officer William Murdoch to First and First Officer Charles Lightoller to Second. It was believed by the officers that "E. J." was retiring after this voyage, and Wilde would probably return to the *Olympic*, and everyone would move back up. There was even talk Wilde might be promoted and given his first command. In any event, the odd man out on this voyage was Second Officer David Blair. Blair, had been with the ship since Belfast, but the Line [translation *"Smith"*] decided that Wilde's experience with her sister was more important. Thus, Lieutenant David "Davy" Blair became the first survivor of the *Titanic*. The missing binoculars were locked in his cabin. It seems he neglected to tell anyone before leaving the ship with the key.

Upon arrival in Southampton, she was berthed at the Ocean Dock [aka The White Star Dock, Piers 43 & 44, built for the *Olympic*-class] and became the center of attention. The process of getting the newest White Star liner ready for sea had began. First up was the lifeblood of any steamship, coal. *RMS Titanic* had a voracious appetite. The week in port, she consumed 415 tons for steam to operate cargo winches and to provide light and heat throughout the ship. The fueling of a ship in the age of coal was a very dirty business. After every ventilator was covered with canvas, air-vent louvers closed and all interior spaces sealed off, eight thousand tons of coal, much of it "on loan" from other IMM ships, were loaded by stokers [known as *"the black gang"*] from lighters [barges] through the side coaling ports. It took 24 hours to re-fuel a large liner like the *Titanic*.

Members of the *"black gang"* after coaling their ship

The Officers of *RMS Titanic*

After which the ship's carpenter would seal up the coaling ports with buckram gaskets soaked in red lead. Finally, each railing, deck, staircase, and passageway had to be cleaned thoroughly to remove the fine coating of black dust that spread seemingly everywhere.

Once the ship's appetite was taken care of, it was time to take care of the passengers, literally. The Victualling Department provided her with provisions in massive quantities, enough food and drink to feed a small town for a month, or in this case, a floating city for a week.

Fresh Meat	75,000 lbs	Potatoes	40 tons
Fresh Fish	11,000 lbs	Onions	3,500 lbs
Poultry & Game	25,000 lbs	Rice, dried beans	5 tons
Salt & dried fish	4,000 lbs	Lettuce	7,000 hd
Bacon & Ham	7,500 lbs	Tomatoes	5,500 lbs
Sausage	2,500 lbs	Fresh green peas	2,250 lbs
Fresh eggs	40,000	Fresh asparagus	800 bun
Flour	200 barrels	Oranges	36,000
Sugar	10,000 lbs	Lemons	16,000
Coffee	2,200 lbs	Grapefruit	50 boxes
Tea	800 lbs	Hot house grapes	1,000 lbs
Cereals	10,000 lbs	Fresh milk	1,500 gal
Ice cream	1,750 qts	Fresh butter	3 tons
Sweetbreads	1,000	Jams	1,120 lbs
Beer, & stout	20,000 bot	Wine	1,500 bot
Mineral water	15,000 bot	Spirits	850 bot

The food was the finest to be found. No vendor in his right mind was going to disappoint a customer who ordered 50 tons of meat once a week. White Star foraged near and far to find the best. Some of the consumables [translation *"wine"*] were sufficient for a round trip.

Chief Purser Hugh McElroy & Captain Edward Smith

The Europeans who would grow the grapes and eventually make quality wine were still arriving [some on the *Titanic*]. That said, the Americans could return the favor on eastbound voyages. Nowhere in Europe was there anything to compare with a New England lobster.

The *Titanic*, like every liner, or for that matter, every cruise ship, was basically a floating hotel. The man who ran the ship was Captain Edward Smith. The man who ran the hotel was Hugh Walter McElroy, the Chief Purser. McElroy had been with Smith on the *Olympic*, and the Captain, knowing the value of a good purser, brought him to his new command.

And a very good purser he was. His blend of Irish charm, persuasion and tact could be brought to bear in any situation. McElroy, like his Captain, knew everyone in First Class, if not personally at least by name. Now imagine staying at a five-star hotel where the general manager knows you, and is there to serve. *"Do you hate your cabin? Go see, Mr. McElroy." "Want to send a Marconigram? Go see, Mr. McElroy." "Want to store your jewels in the safe?"* [Get the idea?] He was immensely popular with the people in First. How popular? The toughest ticket on the *Titanic* was a seat at the Captain's table. The second toughest was a seat at Chief Purser McElroy's table.

On April 9, Thomas Andrews, having finished his work for the day, posted a letter to his wife Helen:

> "The Titanic is now about complete and will, I think, do the old Firm credit tomorrow when we sail."

After the coaling and provisioning were completed, the ship took it's "finals." A representative of the Board of Trade, Captain Maurice Clarke, came aboard to conduct the mandatory surveys and tests.

Board of Trade Inspection, *RMS Titanic*

J. J. Astor about to board the boat train, Paris

Rockets, flares, and other pyrotechnics were examined. The Captain then observed tests of the lifeboats and lifebelts and finally came to the bridge to inspect the charts and instruments. Second Officer Lightoller saw him off at the ladder, not at all sorry to see him depart.

> *"The Board of Trade Surveyor, Captain Clarke, certainly lived up to his reputation of being the best cursed B.O.T. representative in the South of England. Many small details that another surveyor would have taken in his stride, accepting the statement of the officer concerned, was not good enough for Clarke. He must see everything, and himself check every item that concerned the survey. He would not accept anyone's word as sufficient...and got heartily cursed in consequence."*

These remarks from Lightoller, a man with a reputation for being circumspect himself, tells us that Clarke did a through job. The Captain passed her for being in compliance with all particulars. The Board of Trade was satisfied. The lifebelts and pyrotechnics were stowed, and the lifeboats covered. It would be the last time the wooden boats were touched by human hands until shortly after midnight, April 15th. Harland & Wolff's hull No. 401 was, in the words of a future generation, *"good to go."*

Chapter 3
Down to the Sea

"Being in a ship is like being in a jail, with the chance of being drowned."

Samuel Johnson

> **DAILY NEWS**
> APRIL 10, 1912
> **FRENCH LINER NIAGARA STRIKES ICE
> SAILING FROM LE HAVRE TO NEW YORK
> BOW PLATES DAMAGED
> CREW ABLE TO MAKE REPAIRS**

Time: Noon. Date: April 10, 1912. Location: Pier 44, Southampton, England. Departure time for the largest ship in the world to begin her maiden voyage. Second and Third Class passengers had been boarding all morning, with those in Third completing medical exams [the doctors were particularly looking for *Trachoma*, the leading cause of infectious blindness]. White Star took pains to insure that those in steerage were disease-free lest they be refused entry into the United States and then returned to England at the Line's expense.

No such exams were required in First Class. The boat train from Waterloo Station would deliver to the dock the 1912 version of the *"rich and famous."* [The boat train required a First Class ticket.] The *"glitterati"* had arrived, and they boarded last, each to be personally greeted by Captain Smith. E. J. welcomed Isidor and Ida Straus, the owners of *"Macy's"* Department Store. Other "stars" included: Philadelphia streetcar king George Widener, his wife Eleanor, and their son Harry; Banker Robert Daniels; Major Arthur Godfrey Peuchen, President of the Standard Chemical Company; Noël Leslie, the Countess of Rothes; Major Archibald Butt, military aide to Presidents Roosevelt and Taft; and Col. Archibald Gracie, whose family home would one day become the residence of the Mayor of New York. Traveling in Second Class were teachers, merchants, students, priests, servants and members of the ship's orchestra.

RMS Titanic departing Ocean Dock [Pier 44], April 10, 1912

RMS Titanic narrowly avoids the SS New York

Third Class included David *"Dai"* Bowen, a professional boxer and the Welsh Lightweight Champion. Also in First, one would find Joseph Bruce Ismay, the Managing Director of the White Star Line.

What about the bunker fire? Some boiler room survivors claimed the fire had been extinguished in Southampton, while others said it blazed until doused by the fatal influx of seawater four nights later. As for the binoculars, the lookouts were still without field glasses.

Promptly at noon, with one blast from the ship's whistle, the lines were cast off, and the *RMS Titanic* was eased away from the Ocean Dock by tugs. To port lay the Inman Line steamer *New York*, moored along side White Star's own *Oceanic*. As the new ship came abeam, suction was created between them. Steel hawsers were torn from their moorings, and the liner drifted perilously close. A collision was averted when Smith ordered the port propeller reversed, turning the larger liner while a tugboat towed the *New York* in the opposite direction. Neither ship was damaged. Ahead in Solent Roads, Frank Beken waited patiently [in a rowboat] for the liner to pass. Camera at the ready, he would "snap" some of the most famous pictures of the *Titanic* ever taken. Captain Smith [with a grin] ordered four blasts from the whistle in a salute to the intrepid photographer.

At 6:30pm, she arrived at Cherbourg an hour behind schedule. The French Government had felt no compulsion to build a berth big enough to support a British "superliner." As a result, she dropped anchor in the harbor. The passengers who boarded *Titanic* at the French port, did so by tender. Enter the *Nomadic* [and the *Traffic*] to ferry the Line's customers of all classes to the ship. This was the job for which they were built, and Harland & Wolff managed to have each in service in time for the arrival of the *Olympic*.

Father Browne's Photo of passengers, *RMS Titanic*

***RMS Titanic*, outbound from Cherbourg**

The boat train from Paris had arrived at 4pm. On board the first tender were John Jacob Astor IV, one of the richest men in the world, and his second wife, Madeleine Force Astor. The new Mrs. Astor was eighteen years old and four months pregnant. The Astor party included personal servants for himself and Mrs. Astor, as well as a nursemaid to aid with the pregnancy. Next came business titan Benjamin Guggenheim escorting his *paramour*. They were crossing with their retinue, consisting of her maid, his valet and chauffeur. The *Nomadic* ferried landowner Sir Cosmo and Lady Lucy Duff-Gordon. Publisher Henry Sleeper Harper and his wife Myra also came out on the tender, as did John and Marian Thayer and their son Jack. Thayer was a Vice-President of the Pennsylvania Railroad.

Boarding the ship was a couple that had [for good reason] booked passage under assumed names. It would have been an Edwardian *faux pas* for Mr. and Mrs. G. Thorne to admit they weren't married. They were in fact George Rosenshine and Maybelle Thorne, traveling in First Class, posing as man and wife. Quigg Baxter, along with his mother and sister, had cabins on B-Deck. He had also arranged for a cabin on C- Deck for his lover, cabaret singer Bertha Mayné. Also stepping off the tender was the *nouveau riche* wife of a Colorado silver miner, Mrs. Margaret Brown. More Second and Third Class passengers came out on the *Traffic* and twenty-four who boarded in Southampton were disembarked. In the liner's holds could be found a cask of china for *Tiffany & Company*, a custom-made *Renault* automobile, and a jeweled copy of *The Rubáiyát of Omar Khayyám*.

At 8:20pm, the ship weighed anchor and leaving *"La Belle France"* behind, set a course for Southern Ireland. Traversing the English Channel and the Irish Sea she reached *Roche's Point,* the outer anchorage of Queenstown Harbor, at 11.30am on Thursday, April 11.

Third Class Passengers, Queenstown, April 11, 1912

Last picture of *RMS Titanic*, outbound from Queenstown

[Again without a pier] she anchored three thousand yards from shore with the tenders *Ireland* and *America* ready to do the heavy lifting.

Eight passengers disembarked. One of them, the Reverend Francis Browne, was a Jesuit priest. The padre was what future generations would refer to as a *"shutter bug."* While onboard, he took numerous pictures of the new liner. Some of which can be found in this book. Along with them, a member of the *"black gang,"* John Coffey used the opportunity to jump ship in his hometown. The 23-year-old stoker had previously served on the *Olympic*. On Sunday morning he would sign on and sail on yet another liner, the *RMS Mauretania*.

At this stop, a few select vendors were allowed to come aboard and display their wares on the deck. J. J. Astor fancied a lace sports coat, and promptly handed the merchant seven hundred dollars.

The tenders returned, bringing with them 123 new faces for the voyage to America. These passengers were mostly Irish immigrants traveling in steerage [Third Class] to the New World. Among those boarding was a farmer named John Bourke. He was emigrating to the United States, along with his wife Katherine and sister Mary, all of them westbound to Chicago. In Gaelic, it was called *"Gorta Mór,"* "The Great Hunger." The Great Potato Famine would kill a million people and, in the next half-century, send another million away from their homeland, all of them seeking a chance for something better.

Leaving one's native land would be difficult for anyone, but quite possibly hardest of all on the Irish, a people who had suffered terribly, while still grounded in the *"old sod"* and their faith. Now the uncertainty of what was ahead eclipsed the certainty of what was left behind. A folk song of the time expressed the feelings of an adventurous lad who was leaving to seek his fortune in Pennsylvania.

Oh, me name is Paddy Leary
from a spot in Tipperary.
The hearts of all the girls I'm a thorn in.
But come the break of morning
it is they who'll be forlorn.
For I'm off to Philadelphia in the morning.

"With me bundle on me shoulder
sure there's no man can be bolder
I'm leaving just the spot that I was born in.
But some day I'll take the notion
to come back across that ocean.
To me home in dear
old Ireland in the morning,"

Since the *"M"* in *RMS* stands for Mail, the tenders returned with a cargo of a most important type. Fourteen hundred sacks of mail were hoisted aboard and sent down to the mailroom on G-Deck. The *Titanic* had a post office and five clerks, [three employed by the U.S. Post Office] working thirteen hours a day, to sort the mail and to handle any letters that were mailed from the ship by passengers and crew. Many took advantage of the opportunity to post a letter. One of them was Francis Davis Millet, an artist from Massachusetts. From the Liner, Millet wrote to a friend. The letter was mailed in Queenstown. In it, he "comments" about his fellow passengers:

> "Queer lot of people on the ship. There are a number of obnoxious, ostentatious American women, the scourge of any place they infest and worse on shipboard than anywhere. Many of them carry tiny dogs, and lead husbands around like pet lambs."

A somewhat more disturbing letter was posted by the ship's new Chief Officer, Henry Wilde. He wrote his sister:

"I still don't like this ship... I have a queer feeling about it."

By noon the cargo, mail and passengers had been loaded. At 1:30pm, *RMS Titanic* cleared Queenstown, and soon rounded the *Old Head of Kinsale.* She stopped at Daunt's Rock Lightship to drop off John Cotter, the harbor pilot, and then proceeded along the southern coast of Ireland, past *Galley Head* and *Stags,* gathering way toward *Fastnet Light* and then, the North Atlantic. From the foremast she flew the Stars and Stripes to signify her destination country. From the mainmast flew a red pennant with a single white star, the house flag of the White Star Line. And because Captain Smith was a Commander in the Royal Naval Reserve and, could be called to active duty in time of war, she flew the Blue Ensign from the jackstaff. Merchant ships normally fly the Red Ensign. Often referred to by seamen as *"the Red Duster,"* it was [and is] the symbol of the British Merchant Marine. Were the *Titanic* herself *"called to the colors,"* she would hoist the White Ensign, since before the age of Nelson, the iconic standard of the Royal Navy. Next scheduled stop: New York.

Once at sea, life as it always does on a voyage slows down. Westbound, a succession of 25-hour days, [with the time change] leaves plenty of time for those in Third Class to wonder about what would be waiting for them, or to watch their children playing in the small deck area that was allotted to steerage passengers. Second Class had time enough to calculate what they spent and if they had enough money to get home, or find a book in the library on C-Deck. There was time for those in First Class to have a workout in the ship's gym or a dip in the pool. And for all, there were six more slow days until *Titanic* tied up at the newly enlarged Pier 59 on the Hudson.

Morgan Robertson's *Futility*, published 1898

Only surviving photo, *Titanic*'s Wireless Room

In Greek mythology, Cassandra was the daughter of Priam, the King of Troy. The God Apollo, enamored by her beauty, granted her the gift of prophecy. Cassandra instantly spurned his love, and Apollo placed a curse on her so that no one would believe her predictions. It is unlikely that the majority of the 2300 aboard *Titanic* had ever heard the name Morgan Robertson. A case could be made that he was the latter-day Cassandra. In 1898, Robertson, a struggling writer, published a novella. In it, a company built the largest liner in the world. Eight hundred feet long and displacing seventy thousand tons, she was lauded by many as *"unsinkable."* The author loaded the ship with wealthy and famous people and sent her to sea. Four days later [a frigid April night] he wrecked her on an iceberg. His ship was woefully short of lifeboats, and the death toll was enormous. The book was titled *"Futility."* What was the name of the Liner? *Titan*.

Of all the wonders aboard *Titanic*, perhaps the greatest and surely the most important could be found in a pair of cabins above the Boat Deck, adjacent to the officers' quarters. There two young men, John Phillips and Harold Bride, worked one of the two most powerful wireless sets that had ever put to sea. [*Olympic* had the other.] Wireless, the forerunner of radio, was the invention of an Italian scientist, Guglielmo Marconi. A telegraph key would transmit signals in Morse code through the air over great distances. The system was used to send ship-to-ship traffic, house traffic from ship-to-company, and personal messages ship-to-shore. [Personal messages carried *Titanic's* code word, *"ADVISELUM."*] Its most important use, however, was to send and receive warnings and advisories, most often about ice. The set operated into a 4-wire antenna suspended between the masts, some 250 feet above the sea. The equipment was on-board when the liner left Harland & Wolff on her trials.

The Marconi Company guaranteed the equipment's working range was 250 miles, but communications could be maintained for up to four hundred miles during daylight and up to two thousand miles at night. Signals as far away as Tenerife [in the Canary Islands] were being received. A wireless operator ["W/O" for short] was considered a highly skilled employee and was paid accordingly. Second Operator Bride's rate of pay was nine times that of a lookout and sixteen times that of a stewardess. Only Captain Smith had a higher pay grade.

First Class cabin A36 was, for the round-trip voyage, assigned to the builder's representative, Thomas Andrews. Piled in his stateroom [which did not appear in the ship's plans] were charts, diagrams, and blueprints of the *Titanic*. Andrews spent his days touring the ship, looking for problems, taking notes, and suggesting improvements. In the whole world, no one knew as much about the *Olympic*-class liners as he did, not even Carlisle and Pirrie, the original designers.

He had been with Harland & Wolff since he was sixteen. As an apprentice, he began with three months in the joiners' shop, followed by a month in the cabinetmakers, and then a further two months working on the ships. The last eighteen months of his five-year apprenticeship were spent in the drawing office.

In 1901, Andrews, after working his way up through the many departments of the company, became the manager of the construction works. That same year, he became a member of the Institution of Naval Architects. In 1907, Andrews succeeded Carlisle and became Managing Director and head of the drafting department at Harland & Wolff. His chief project would be the two leviathans that the yard was about to build. To prepare for this assignment, he learned everything that could be learned about the ships.

Thomas Andrews, Managing Director, Harland & Wolff

Helen, "ELBA" and Thomas Andrews

No detail escaped his grasp. In his biography, *"Thomas Andrews, Shipbuilder,"* [written and published shortly after the wreck] Shan Bullock paints the portrait of an exceptional man, as only a fellow Irishman could tell it.

> *"One sees him, big and strong, a paint-smeared bowler hat on his crown, grease on his boots and the pockets of his blue jacket stuffed with plans, now consulting his Chief, now conferring with a foreman, now interviewing an owner, now poring over intricate calculations in the Drawing office, now in company with his warm friend, old schoolfellow, and co-director, Mr. George Cumming of the Engineering department, superintending the hoisting of a boiler by the two hundred ton crane into some newly launched ship by a wharf.*
>
> *Or he runs amok through a gang--to their admiration, or comes unawares upon a party enjoying a stolen smoke below a tunnel-shaft, and, having spoken his mind forcibly, accepts with a smile the dismayed sentinel's excuse that 'twasn't fair to catch him by coming like that into the tunnel instead of by the way he was expected."*
>
> *"Or he kicks a red-hot rivet, which has fallen fifty feet from an upper deck, missing his head by inches, and strides on laughing at his escape. Or he calls some laggard to stern account, promising him the gate double quick without any talk next time. Or he lends a ready hand to one in difficulties; or just in time saves another from falling down a hold; or saying that married men's lives are precious, orders back a third from some dangerous place and himself*

> takes the risk. Or at horn-blow he stands by a ship's gangway, down which four thousand hungry men, with a ninety feet drop below them, are rushing for home and supper, and with voice and eye controls them ... a guard rope breaks . . . another instant and there may be grim panic on the gangway . . . but his great voice rings out, "Stand back, men," and he holds them as in a leash until the rope is made good again."

Andrews was as good with people as he was with ships, liked by everyone and always available for those with whom he worked. Officers, Engineers, and Staff were forever coming to him with their problems. When *Titanic* sailed from Belfast, he left behind his wife Helen and their year-old daughter Elizabeth Law Barber Andrews, known by her initials, "ELBA." He was thirty-nine years old.

The *Titanic* was under the command of Captain Edward John Smith. Smith, often referred to by his men as "E. J.," joined The White Star Line in March 1880 as the Fourth Officer of the *RMS Celtic*. He served aboard the Company's liners to Australia and to New York, where he quickly rose in status. In 1887, he received his first White Star command, a Western Ocean Mail boat, the *RMS Republic*.

The following year Smith earned his Extra Master's Certificate. He was *RMS Majestic's* captain for nine years commencing in 1895. When the Boer War started in 1899, *Majestic* was called upon to transport troops to Cape Colony. Smith made two trips to South Africa, both without incident, and, for his service, King Edward VII awarded him the *"Transport Medal"* in 1903. He was regarded as a "safe captain." As he rose in seniority, he gained a reputation amongst passengers and crew for *"quiet flamboyance."*

Captain Edward J. Smith, Commander, *RMS Titanic*

The Captain was known for his calm demeanor, but on those rare occasions when he did raise his voice, men would jump. Some passengers [especially First Class regulars] would sail the Atlantic only in a ship he commanded. From 1904 on, the man known as the *"Millionaire's Captain"* was master of the White Star Line's newest ships on their maiden voyages.

In 1904, he was given command of the then largest ship in the world, the *RMS Baltic*. Her maiden voyage from Liverpool to New York in June went without incident. After three years with the *Baltic*, Smith was given his second "big ship," the *RMS Adriatic*. Once again the first sailing came off without a hitch. In 1907, while in command of the *Adriatic*, Smith granted an interview to a reporter. His answer to the question of what he feared at sea was eerily prescient.

> *"We do not care anything for the heaviest storms in these big ships. It is fog that we fear. The big icebergs that drift into warmer water melt much more rapidly under water than on the surface, and sometimes a sharp, low reef extending two or three hundred feet beneath the sea is formed. If a vessel should run on one of these reefs half her bottom might be torn away."*

In response to anoher question from the reporter, he summarized his life at sea:

> *"When anyone asks how I can best describe my experience in nearly 40 years at sea, I merely say, uneventful. Of course there have been winter gales, and storms and fog the like, but in all my experience, I have never been in any accident of any sort worth speaking about. I never saw a wreck and never have been wrecked, nor was I ever in any*

Lord Pirrie and Captain Smith, Boat Deck, *RMS Titanic*

> *predicament that threatened to end in disaster of any sort. You see, I am not very good material for a story."*

[Later the Captain would add his entry in the *"famous last words"* Hall of Fame.]

> *"I cannot imagine any condition which would cause a ship to founder. I cannot conceive of any vital disaster happening to this vessel. Modern ship building has gone beyond that."*

Smith had built a reputation as one of the world's most experienced sea captains and so was called upon to be the first Commander of the lead ship in the Line's latest class of ocean liners, the *RMS Olympic,* new holder of the crown, *"the largest ship in the world."* She departed Southampton on her maiden voyage to New York on June 14, 1911. A week later, she reached Manhattan without an indident following stops in France and Ireland. Her good luck would hold for three months.

On 20 September 1911, *Olympic's* first major mishap occurred. While passing through the Solent, she collided with a British cruiser, *HMS Hawke.* [Smith now had an accident worth talking about.] Although the collision left a huge gash in her starboard side shell plating and one of her propeller shafts twisted, she was able to limp home to Southampton and then onto Harland & Wolff to make repairs. At the Inquiry, the Board of Trade blamed the *Olympic* for the incident, alleging that her massive size generated a suction that pulled the *Hawke* into her side. The fact that the cruiser's steering gear was jammed at the time did not seem important to the Board. The incident occurred with Captain Smith on the bridge. *The New York Times* believed that soon there would be a new Commodore.

> "Capt. E. J. Smith, R. N. R., the Commodore of the White Star Line, who is to command the new mammoth liner Olympic, will retire at the end of the present year, it is understood, as he will have reached the age limit. He will be relieved by Capt. H. J. Haddock of the Oceanic."

Yes, Captain Smith had indeed reached retirement age, but no, he wasn't quite ready to call it a career.

> "The second big liner, the Titanic, which is to enter the New York to Southampton service toward the end of the year, will be commanded, it is said, either by Capt. B. H. Hayes of the Adriatic or Capt. Henry Smith."

The *Times* got this one wrong. Haddock did get command of the *Olympic*, but Bertram Hayes and Henry Smith were not moving up. [Hayes would, however, eventually command the *Olympic*.] The *Hawke* incident was a financial disaster for White Star, exacerbated by the out-of-service time for the big liner. Despite the past trouble, Smith's confidence in the new ships remained high, as did the Line's faith in him. He was set to Captain the newest *Olympic*-class liner.

> "The Olympic is unsinkable, and Titanic will be the same when she is put in commission. Either of these two vessels could be cut in halves and each half would remain afloat almost indefinitely. The non-sinkable vessel has been reached in these two wonderful craft. I venture to add that even if the engines and boilers of these vessels were to fall through their bottoms, the vessels would remain afloat."

The politics of the White Star Line were not lost on the Captain. Smith had memorized the party line and downed the Kool-Aid.

RMS Britannic (Formally *RMS Gigantic*) on the stocks

Although several sources stated that the Captain had decided to retire after completing *Titanic*'s first sailing, an article in the *Halifax Morning Chronicle* on 9 April 1912, affirmed that Smith would indeed, "*remain in charge of Titanic, until the Company completed a larger and finer steamer.*" [Presumably the *Gigantic*].

Titanic's plans called for the forward section of the Promenade Deck to be enclosed. The revision would change both the interior and exterior. No more was she the mirror image of the *Olympic*. As a result of the B-Deck windows, the twins were no longer identical. The modifications also created the four Parlor Suites, the most expensive accommodations aboard. These spaces contained two bedrooms, a sitting room, a private bath, and a private promenade. No other ship in the world could offer personal deck space, not even the *Olympic*. The cost [in 2016 dollars] was upwards of $80,000. One of the Suites, B52/54/56, already had a back story. In early February, the American steel baron Henry Clay Frick reserved the rooms but cancelled when his wife sprained her ankle. The booking then passed to J. P. Morgan, who gave it up, choosing to remain at a resort in Aix-les-Bains, France. [Morgan's reprieve was short-lived; a year later he died in his sleep.] Yet another wealthy American, J. Horace Harding took over the reservation but at the last minute changed his mind and sailed instead on the *Mauretania*. No one else reserved the rooms, so on her maiden voyage, *Titanic's* Parlor Suite B52/54/56 was occupied by a man who was traveling on the house. His name was Ismay.

Joseph Bruce Ismay was born in Crosby, Lancashire, a small town near Liverpool. He was the son of Thomas Ismay, senior partner in Ismay, Imrie and Company and founder of the White Star Line. It was known from the start that his future would be with the firm.

J. Bruce Ismay, Managing Director, White Star Line

Parlor Suite B52/54/56 Promenade, *RMS Titanic*

Alfred Gwynne Vanderbilt

Educated at Harrow, he spent a year touring the world, and then went to New York City as the company's representative. In December 1888, Ismay married Julia Florence Schieffelin, daughter of prominent attorney and art collector, George Richard Schieffelin. The following year Thomas Ismay died, and Joseph became head of the family business. And like many a son before him, his first decision was to build four liners that surpassed the *Oceanic* built by his father. These vessels were designed more for luxury and safety than for speed. They also brought White Star into direct competition with England's leading shipping company, Cunard. In 1901, he was approached by a group of Americans, headed by J. P. Morgan, who wished to create a transatlantic shipping conglomerate. He subsequently agreed to merge his firm into the International Mercantile Marine. Ismay, as was his custom, accompanied his ships on their maiden voyage. Thus, it was no surprise to see him on the first sailing of The White Star Line's greatest achievement.

Titanic was fabled for those aboard and those who *"missed the boat."* If David Blair was the ship's first survivor, the second might well be Alfred Gwynne Vanderbilt [of the famous and wealthy Vanderbilts]. In April 1912, he and his wife were in Europe and scheduled to return home on the *Titanic*. Family members warned him about potential problems on a maiden voyage. So, on April 9, the day before the *Titanic* sailed, Vanderbilt cancelled the booking. A servant, Frederick Wheeler did make the voyage with their luggage and perished. Three years later Alfred's luck would run out. On a trip to Europe he was aboard a ship that was torpedoed off the coast of Ireland. Several survivors reported they last saw him offering his life vest to a child and then helping the mother tie it onto the infant. Vanderbilt did not survive. The liner's name was the *RMS Lusitania*.

Chapter 4
April Crossing

"Ships are the nearest things to dreams
That hands have ever made."

Robert N. Rose

> "When the cabin port-holes are dark and green
> Because of the seas outside;
> When the ship goes wop (with a wiggle between)
> And the steward falls into the soup-tureen,
> And the trunks begin to slide,
> When Nursey lies on the floor in a heap,
> And mummy tells you to let her sleep,
> And you aren't waked or washed or dressed,
> Why then you will know (if you haven't guessed)
> You're Fifty North and Forty West!"
>
> *Rudyard Kipling*

The North Atlantic Ocean is famous [or infamous] for having some of the nastiest weather on the high seas. Storms, high waves, and fog were all too common on the runs from the Continent to New York and Boston and with them all came the dreaded *"mal de mer."* RMS *Titanic* was, at least weather wise, fortunate. During the entire crossing, the weather was clear with the single exception of one small patch of fog, and the sea was remarkably calm throughout the voyage. There was sunshine the whole of each day and bright starlight, albeit with cold temperatures, every night.

The ship was settling into it's routine. First Class passengers could avail themselves of a myriad of creature comforts, starting with the ship's gymnasium. In spite of its namesake's girth, Edwardian England had an affinity for physical fitness. The *Titanic's* gym featured an electric camel, electric horses, cycling machines, and a rowing machine. In addition, a fitness instructor, Thomas W. McCawley [replete in white flannels] stood by to assist. Tickets, priced at one shilling, were available from Purser McElroy and entitled First Class passengers to one session in the facility.

Thomas W. McCawley, Fitness Instructor. *RMS Titanic*

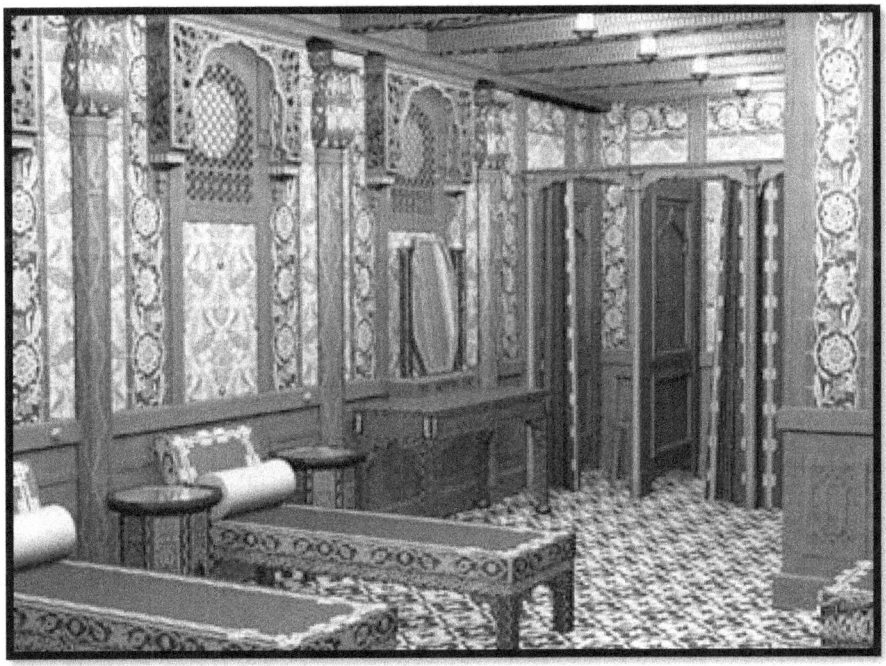
Turkish Bath, *RMS Titanic*

The gymnasium was open for ladies between 9.00 am and noon. It was reserved for gentlemen between 2.00pm and 6.00pm. Children were allowed in between 1.00pm and 2.00pm. It was here, according to a White Star pamphlet that:

> *"Passengers can indulge in the action of horse riding, cycling, boat rowing etc, and obtain beneficial exercise, besides endless amusement."*

After an hour in the gym, one might want to take advantage of another feature. For a charge of four shillings [about a dollar], First Class passengers could soothe away their aches and pains at the Turkish Baths. The suite had a steam room, a hot room, a temperate room, shampooing rooms, toilets, and a cooling room. There was also an ultra-modern innovation: electric beds that applied heat to the body using electric lamps.

How about a little more exercise? The squash racquet court provided for First Class passengers was located down on G-Deck. It was under the supervision of *"Racquet Professional"* [Frederick Wright] who supplied equipment and who would play as an opponent if required. Players were charged two shillings [50 cents] for the use of the court, and games were limited to one hour if others were waiting. The squash court compartment included an observers' gallery on F-Deck. The gallery would serve in an unintended role on Sunday night.

And, if all that wasn't enough, White Star had one more exclusive it could offer to the *crème* of Atlantic travelers. Taking advantage of the abundance of space in their new liners, the Line gave the *Olympic* and *Titanic* the first heated [salt water] swimming pools ever installed in a ship. This last little extra was enough to put it over the top for the *Belfast Evening Telegraph:*

Swimming Pool, *RMS Titanic*

First Class Dining Salon, D-Deck, *RMS Titanic*

> *"Then the morning plunge in the great swimming bath, where the ceaseless ripple of the tepid sea water was almost the only indication that somewhere in the distance 72,000 horses in the guise of steam engines fretted and strained under the skilful guidance of the engineers."*

For transatlantic passengers, intense interest in two things have remained the same from the mid-nineteenth century and the first *Britannic* to the dawn of the twenty-first century and the *Queen Mary 2*. One would be accommodations and the other, food. Simply put, *"where am I sleeping and what am I eating?"* We'll start with food. The White Star Line, having outstripped its competition in luxury, was not about to be out done in the commissary department.

The *Titanic's* largest public room was the First Class Dining Salon. The Jacobean-style room, located on D-Deck between the second and third funnels, extended the full width of the ship. Nearly six hundred passengers [virtually all of First Class] could be handled in one sitting. As noted, the room was located at the center of the ship to give First Class diners the smoothest ride possible [with the rest of the vessel, in essence, built around this room]. The floor of the Salon was laid with linoleum tiles intricately patterned to resemble a Persian carpet. The small tables made for easy conversation between tablemates, an activity no doubt assisted by the *haute cuisine*, fine wine, and comfortable armchairs. The mood was further enhanced by the presence of Wallace Hartley and the ship's orchestra.

> *"This immense room has been decorated in a style peculiarly English that, in fact, which was evolved by the eminent architects of early Jacobean times. The furniture of oak is designed to harmonize with its surroundings."*

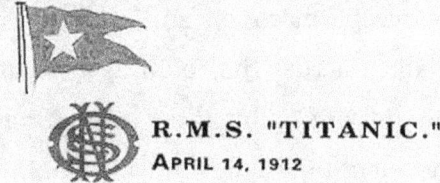

R.M.S. "TITANIC."
APRIL 14, 1912

FIRST CLASS DINNER.

Hors d'Oeuvre Variès

Oysters

Consomme Olga   Cream of Barley

Salmon, Mousseline Sauce, Cucumber

Filet Mignons Lili

Sauté of Chicken Lyonnaise

Vegetable Marrow Farcie

Lamb, Mint Sauce

Roast Duckling, Apple Sauce

Sirloin of Beef Chateau Potatoes

Green Peas   Creamed Carrots

Boiled Rice

Parmentier & Boiled New Potatoes

Punch Romaine

Roast Squab & Cress

Red Burgundy

Cold Asparagus   Vinaigrette

Pâté de Foie Gras

Celery

Waldorf Pudding

Peaches in Chartreuse Jelly

Chocolate & Vanilla Eclairs

French Ice Cream

First Class Dinner Menu, April 14, 1912, *RMS Titanic*

Dinner was served at 6pm. First Class passengers made an entrance from the reception room, and clad in their jewels and *"tres chic"* finery, walked across the room to a waiting table and stewards. In the center of the Salon, Edward J. Smith dined with a carefully chosen dozen of their fellow travelers at the Captain's table.

For those not satisfied with the First Class Dining Salon, there was another choice. Around the turn of the century, a visionary named Albert Ballin had become head of Germany's Hamburg-America Line. His first ship, the *Amerika*, entered service in 1905 and was at the time, the largest steamer in the world. Built by Harland & Wolff [they couldn't build for Cunard, but everyone else was fair game], she featured two things never before seen on the North Atlantic, an elevator, and an *á la carte* Restaurant. Ballin had cut a deal with Ritz-Carlton to operate an alternative dining room on his ship. A separate staff, trained personally by César Ritz, combined with their own galley and chefs schooled by Auguste Escoffier, available 24 hours a day. All set in a sumptuous Salon, designed by Charles Mewès, the architect of *The Hôtel Ritz* in Paris and *The Ritz Hotel* in London. Ballin believed *"The Gilded Age"* was ready for this and he was soo right. It was a smashing success. In the war for supremacy on the high seas, his ship might be losing in the engine room, but she was running away in the dining room. Until 1907, and the arrival of *RMS Lusitania,* she was for savvy travelers, *"the only way to cross."*

[*"If you can't innovate, imitate."*] Yet again, White Star would have to play catch up. When the *Olympic* put to sea in 1911, she did so with a separate 25 table restaurant, located aft on the bridge deck. So popular was the *á la carte* that modifications were made in *Titanic*. Her restaurant was expanded to hold 140 First Class diners.

The *á la carte* Restaurant, *SS America*

The *á la carte* Restaurant, *RMS Titanic*

In the process, the *Titanic* and the other liners of her time had done a little class engineering. The addition of this exclusive and expensive dining room helped create an *"Über Class"*, somewhere above First Class. A reservation in the *á la carte* brought a certain "aloofness" to the mealtime experience, if for no other reason than you were dining two decks above First Class. It was also the perfect location to *fête* friends at an elegant private dinner party.

Now just being special was not good enough. Meals for all classes on *Titanic* were included in the price of the passage, and diners were constrained by a fixed menu. However, in the *á la Carte* Restaurant, diners could chose each course separately from a wider selection than that available in the main Dining Salon. It was akin to the modern day First Class dining rooms on the *QE2* and *QM2*. If a passenger asks for something that isn't on the menu, a waiter is most likely to respond, *"Sir [or Ma'am], if it's on the boat, you can have it."* One passenger summed up the whole experience.

> *"Fancy strawberries in April, and in mid ocean. The whole thing is positively uncanny. Why, you would think you were at the Ritz."*

And some did indeed refer to this Dining Salon as *"The Ritz."* It was as if you were dining in the finest restaurant on shore, with all the "trimmings" - right down to the check. Passengers had to pay for their food out of pocket and were presented with a bill at the conclusion of their meal. White Star would take from £3 to £5 off the price of their passage ticket so long as the traveler made exclusive use of the *á la Carte* Restaurant. One wag said of trying to cross the Atlantic with only five quid to spend in the Restaurant, *"One might feel just a bit peckish by the time they got to New York."*

The *"Café Parisien,"* Bridge Deck, *RMS Titanic*

Second class Dining Room, D-Deck, *RMS Titanic*

If one is good, then two must be better. Adjacent to the *á la Carte* was yet another restaurant, the *Café Parisien*. The *Café* offered the same menu as the *á la Carte*, in a somewhat different, less formal [rattan/wicker] setting. In the *Café*, large picture windows afforded a view of the sea while dining, previously unheard of on a British Ship.

Few things in this world are as malevolent as the weather on the North Atlantic, but, when the conditions were right, the windows could be rolled down and passengers could dine *al fresco*. This was another first for White Star. The Line believed that this space would be a nice retreat for the "children" [twenty years and above] of their First Class passengers, and so it was. On her first and only voyage, the *Café Parisien* proved to be quite popular with *"the younger set."*

Since *The Ritz* was already contracted to their rivals in Hamburg, White Star would have to look elsewhere for its management. Both the *á la Carte* and the *Café* were owned and operated as private concessions by Gaspare Antonio Pietro Gatti. An Italian immigrant to England and a well-regarded London restaurateur, Gatti had his own staff of chefs, waiters, and kitchen help whom he paid from his own pocket. They were not employees of the White Star Line.

As on all White Star Liners, the call to meals would be signaled by a bugler passing from deck to deck. Peter W. Fletcher was his name and he would play before all meals [although to some Americans it sounded less like *"come and get it"* and more like a cavalry charge]. Second Class travelers would proceed to the dining room located on D-Deck aft of the fourth funnel. The large, elegant room could accommodate all five hundred and sixty Second Class passengers in a single seating. The Salon featured mahogany furniture upholstered in crimson, while a piano provided entertainment for the diners.

And while First Class passengers enjoyed the luxury of small tables and moveable chairs, Second Class made do with the traditional long tables and swivel chairs bolted to the floor. This was a precaution against the inevitable storm at sea. A hearty three-course meal with a starter [on April 14th it was consommé], four choices of entrees and seven kinds of desserts were enough to handle every palate. And since the meals were cooked in the same galley as First Class, on the same stoves, using the same ingredients, the quality was excellent.

To the shipping lines, the most lucrative portion of the traveling public was neither First nor Second Class. White Star, like all the other companies running the *"Atlantic Ferry,"* knew that the highest profit margin was to be made further down in the ship. More people had immigrated to the United States and Canada in 1907 than in any year in history. The passing of "refugees" from the Old World to the New was now big business. In 1921, Congress would finally act to cut off the flow, but in 1912 those *"huddled masses"* were money in the bank. The *Titanic* was equipped to house and feed upwards of nine hundred in its lowest class of service. What had once [and sometimes still] known as *"steerage"* because of the livestock formally transported in these spaces was now called Third Class.

There were two Third Class Dining Rooms on F-Deck, two decks below First and Second Class, separated by a watertight partition. [No problem running a bulkhead through their dining room.] They had a capacity of 473. Meals were served in two sittings. The food was simple but substantial. [No one went hungry on the *Titanic*.] The rooms were not decorated in the grandeur of the upper class Salons, but were bright, white, and immaculately clean. White Star offices all over Europe spun the story, neglecting to mention that *"amidships on the middle deck"* was two decks above the engines.

Third Class Dining Room, F-Deck, *RMS Titanic*

First Class Cabin, B-Deck, *RMS Titanic*

> *"Situated amidships on the middle deck, consisting of two saloons extending from ship's side to ship's side, well lighted with sidelights, and all finished enamel white; the chairs are of special design. The position of this apartment – in the centre of the ship – illustrates the wonderful strides made in passenger comfort in modern times."*

There was another "Salon" for passengers. In the ship's blueprints it is labeled as the *"maid and valet dining room."* Located on C-Deck, around the corner from the barbershop, it was here that personal servants of First Class passengers took their meals.

If you remember, *"Shipbuilder"* wrote on completion of the *Olympic* that, *"the excellent result defies improvement."* In point of fact, the *Titanic* was improved - for starters, she was four inches longer. A second restaurant, private promenades, a larger reception room, more [and thicker] carpeting along with technical upgrades were among the revisions to her plans. [Andrews missed nothing.] She would become known as *"The Olympic perfected."* First Class cabins were equipped with telephones, heaters, table fans, sinks and call bells for summoning the steward. Private bathrooms, however, were only found in the Parlor Suites. The rooms were rather spacious. To any seasoned traveler, space is the ultimate luxury. [One must have room enough to accommodate that most necessary of traveling companions, the steamer trunk]. All this in spite of the notion that cabins were principally for sleeping or dressing, with other time spent in public rooms or weather permitting on deck.

Second Class was available in one, two, and four berth cabins. All but the single-person rooms would have bunk beds screwed to the wall. In total, the liner had space for 550 passengers in her Middle Class.

The rooms were decorated in enamel white with mahogany furniture. No Middle Class cabin had a private toilet, yet all were provided with sinks and mirrors. Stewards made sure the linen was changed daily. In short, Second Class rooms on the *Titanic* were far ahead of the competition, a fact not lost on the Line.

> *"White Star Line has done much to increase the attractions of second-class accommodation during recent years, having made a special feature of this in a number of their vessels; and in the Olympic and Titanic it will be found that this class of passenger has been generously provided for."*

Much has been made about Second Class on the *Olympic* and the *Titanic*, about how the whole experience was equal to First Class on other liners. The reality was far different. Yes, the rooms might be a little larger, not because Ismay worried about his poorer passengers [believe me, he didn't], but rather because a ship this size had room to spare. Sure, use a little in Middle Class. As the cabins on B-Deck grew larger, so did the ones on F-Deck [ever so slightly]. Regardless, with a swimming pool, gym, restaurants, and service above reproach, it would be impossible to mistake Second for First on the *Titanic*.

At the height of the 19th century, tickets purchased by emigrants migrating to the United States and other nations in the Western Hemisphere only promised *"conveyance"* to the New World. Food, bedding and utensils were seldom included in the price of passage and even the matter of a berth to sleep in was never guaranteed. Aboard the *RMS Titanic* this was surely not the case. Third Class passengers had a place to sleep, in a modest cabin, six to a room. The elimination of open berths forward for single men was another improvement over the *Olympic*.

Representation of a Second Class cabin, *RMS Titanic*

Representation of a Third Class cabin, *RMS Titanic*

The rooms "featured" bunk beds and very little space. Lower Class berths did offer all the *"little luxuries"* that had helped build the reputation of The White Star Line. They provided electric lighting, heat, and washbasins. In fact, the accommodations in Third Class were, for many, far better than the life they left behind.

Third Class living spaces were located on E, F, and G-Decks forward and D, E, F, and G-Decks aft. All were well down in the ship, and those aft close to the engine rooms and the boilers. As in Second, there were no private bathrooms, and only two bathtubs available for all 710 third-class passengers, one each for the gentlemen and the ladies. This was not an oversight on the part on the Line because many people at that time believed that taking frequent baths would lead to lung disease. Edwardian propriety dictated that women and children were isolated aft in the ship. Single men, as noted, were berthed forward, as far away as possible from the females.

The Third Class General Room was a gathering place for the multitude of steerage passengers steaming towards a new life in North America. Located starboard side on C-Deck at the stern of the ship, it served as a lounge, a nursery and a recreational area. The room was finished in white enameled pine and fitted with slat-seated benches and teak chairs. In the interest of hygiene, there were no upholstered surfaces. However, the walls were decorated with posters and pictures advertising the Line's vessels and ports of call. The newspapers decided it was an "upbeat" entrance to a new life.

> *"It is paneled and framed in pine and finished enamel white, with furniture of teak. This will be the general rendezvous of the third-class passengers – men, women and children – and will doubtless prove one of the liveliest*

First Class Smoking Room, A-Deck, *RMS Titanic*

Third Class General Room, *RMS Titanic*

> *rooms on the ship... The new field of endeavor is looked forward to with hope and confidence... the interval between the old life and the new is spent under the happiest possible conditions."*

The popularity of Third Class on White Star's ships continued to grow, aided by of all people, the United States Government. An American agent, Anna Herkner disguised herself as a Bohemian immigrant and made three trans-Atlantic crossings on ships of different lines in order to carry out [in secret] her investigation of the conditions in steerage. She crossed on the German Liners *Friedrich Der Grosse*, and *Pennsylvania* and was appalled by what she found. Stewards sexually assaulting female steerage passengers, a lack of medical care and terrible food. She then crossed on White Star's *Cedric* and was delighted. She shared a clean, comfortable, and private cabin with three other ladies. A steward could be summoned with a call button and the food and service were far better than what Herkner found on the German Ships. Congress would chip in with more help for *"the wretched refuse of your teeming shore."* Third Class was now covered by American Immigration Laws, which mandated a twenty percent increase in space for each passenger, thus, superseding the requirements of the British Board of Trade.

Following dinner, the gentlemen retired to the smoking room, while their ladies were banished to the reading and writing room. The vote for women was still almost a decade away, and Edwardian tradition dictated that the male of the species must have a quiet place to enjoy their brandy and cigars. White Star provided them a large room with a bar, card tables and over-stuffed chairs, just perfect for a little good fellowship. In years to come their wives would storm their way in, but now it was still free of the *"incessant chitchat of women."*

The strict segregation between the social classes, and following the evening meal, between the sexes, continued down to B-Deck and the Second Class smoking room.

And while the room was not quite as plush as the one on A-Deck, it was still rich with Testosterone. The Smoking Room was paneled in carved oak. The oak furniture upholstered in dark green Moroccan leather, added to its masculine air. There was also a Third Class smoking room, not as luxurious but complete with bar and spittoons.

Titanic also featured lounges in First and Second, two libraries, and for good measure a palm court [also known as *The Veranda Café*]. Rooms for everything, except, a place for men and women to be together after dinner. The lounges were for daytime socializing. At night, the ladies went one way, while the gentlemen went the other. Did you want to see you wife after supper and do a little dancing? No problem, just book passage on a German ship. What dancing there was took place in Third Class, where there was always some fellow with a fiddle and another with a squeezebox, playing a jig.

There was something for everyone on the ship. Good food, good rooms, good friends, and in the morning a deck chair with a steamer rug and a cup of bouillon. Plus, there was always a friendly and helpful steward within earshot, for whatever else you required.

On her first day at sea, the *Titanic* covered 386 miles, for an average of 16 knots, [including the time spent at Queenstown]. The patient White Star engineers, as always with new machinery, allowed the expansion engines and the turbine to "warm up." All but five of her twenty-nine boilers were on-line. By the end of the day the new liner was making 21 knots. She was already *"a long way from Tipperary."*

Boarding Pass, *RMS Titanic*

Chapter 5
Shipmates

"I would not creep along the coast
But steer out in mid-sea,
By guidance of the stars."

TS Eliot

In 1912, the Commander of an express liner had two responsibilities: 1) Guide his ship safely to port, and 2) interact socially with the passengers. E. J. Smith was an acknowledged master of both. Now when we say passengers, we mean of course, First Class passengers. No one from Second ever ate at the Captain's table, and I doubt that anyone in Third could pick him out of a mug book. He was still Captain of all, and on this voyage "all" was a rather interesting group.

Their Cabin numbers were C62/64. Not the best accommodations on *Titanic*, but good enough for the richest man on the ship. John Jacob Astor IV and his bride Madeleine Force Astor were returning from their honeymoon. John Jacob IV was the great-grandson of John Jacob Astor whose fortune, made in the fur trade and real estate, helped make the family one of the wealthiest in the United States. His personal wealth would also come from real estate. Astor built the Astoria Hotel, *"the world's most luxurious hotel"* in New York City, adjoining the Waldorf Hotel owned by Astor's cousin, William Waldorf Astoria. The complex would become known as [what else?] the *"Waldorf-Astoria"* [and would serve as the New York host location for the Senate's Inquiry into the loss of the *RMS Titanic*].

In 1891, Astor married Ava Lowie Willing and had two children. In 1909 the couple divorced. The following year at his Bar Harbor home he met a seventeen-year-old girl named Madeleine Talmedge Force and became instantly smitten. Madeleine and her mother, known in society as *"La Force Majeure,"* were soon frequent guests at Astor's homes and estates. Mother Force moved quickly. By August 3rd, her daughter was engaged, and married on September 9, 1911. In the high society of the time divorce was frowned on and re-marriage [especially to a teenager] was anathema. With the élite summering in Newport, Astor went on a *"charm offensive,"* to *"schmooze"* his peers.

John Jacob Astor, First Class passenger, *RMS Titanic*

Madeleine and John Jacob Astor leaving New York

> *"Although practically unknown to the Newport set the future Mrs. Astor was assured an enthusiastic welcome when Mrs. Ogden Mills, social arbiter of the seaside colony, openly expressed her warm approval of the engagement."*

Ah, but Astors lived in Manhattan, and not Rhode Island. In his hometown, J. J. Astor was now *"persona non grata."* Hounded by the tabloids and knowing they would not be regarded as pariahs in Europe, they fled America, *"until the heat died down."* He and the new Mrs. Astor sailed east [on the *RMS Olympic*] and honeymooned on the Continent. When Madeleine became pregnant the couple chose to return home. Their entourage included his valet Victor Robbins, a maid, nurse and Astor's beloved Airedale, *"Kitty."*

One deck above the Astor's was the suite [B82/84] of Benjamin Guggenheim. Like Astor, he was a wealthy businessman. And like Astor he made his money the old fashioned way - he inherited it. Ben was born in Philadelphia, one of seven sons of the wealthy mining magnate Meyer Guggenheim. While Guggenheim inherited a great deal of money from his father, it came without his dad's business acumen. As his losses increased he grew distant from his wife Florette and, ostensibly for "business" reasons, was frequently away from their home. His apartment in Paris was the destination, for a little rest & recreation. Unlike Astor he did not get a divorce but did what a gentleman of his standing would do, he took a mistress. A French singer named Léontine Aubart was now his companion. They planned on sailing back to America on the *Lusitania,* but the Cunarder was laid up for a re-fit at Liverpool. The next westbound steamer was the *Titanic* and they joined her in Cherbourg with their servants, but were discreet enough to stay in separate staterooms. [We'll assume the Mrs. wasn't planning to meet the Liner at Pier 59.]

Benjamin Guggenheim, First Class passenger, *RMS Titanic*

Isidor and Ida Straus, First Class passengers, *RMS Titanic*

In Cabins C55/57 could be found two of New York's most prominent residents, and while most people knew their name, everyone knew their business. The couple was Isidor and Ida Straus, the owners of *Macy's* department store in Manhattan. He was born in Germany in 1845 and came to America nine years later. In 1866 the Straus family moved from Georgia to New York City, where Isidor and his brother set up their family business in *R. H. Macy and Company*. In 1871, Isidor married Rosalie Ida Blun. In 1895, he was elected to Congress and the following year became the owner of the department store. They came to Europe in 1912 on the *Amerika*. As a rule they preferred to travel on German steamers, but booked their return passage on the *Titanic*. As they stepped off the boat train and onto the Ocean Dock in Southampton, he was 67 years-old and Ida was 63. They had been married for forty-one years.

Across the deck in suite C80/82 were the Wideners. George Dunton Widener, like so many others in First Class, was born into money and in his case, tons of the stuff. The eldest son of Peter A. B. Widener, the streetcar king of Philadelphia, he joined his father's business and eventually took over the running of *Philadelphia Traction*. Through hard work and enterprise he became wealthy in his own right. In 1883, he married Eleanor Elkins, the daughter of his father's business partner. They lived at Lynnewood Hall, his father's 110-room Georgian-style mansion. The couple had three children, Harry, George and Eleanor. Harry was with his parents when they left for France, in 1912. The Widener's had just opened a new hotel in Philadelphia [*Ritz-Carlton*] and were looking for a chef. It was no accident that they came home on the *Titanic*. George was also a member of the board of the Fidelity Trust Company of Philadelphia, the firm that bankrolled IMM, the owners of The White Star Line.

George, Eleanor and Harry Widener, *RMS Titanic*

Margaret Brown, First Class passenger, *RMS Titanic*

It was on the afternoon of April 14th, that Widener and his wife watched as Captain Smith handed Ismay a message received from the *RMS Baltic*. Ismay simply glanced at the message, put it in his pocket and headed below. That evening Smith was the guest of honor at a dinner party given by George and Eleanor in the *á la carte* Restaurant. Following dinner, per custom, the gentlemen retired to the smoking room. Four of their guests were still there at 11:40pm.

Margaret Tobin was born in Hannibal, Missouri in 1867. Her parents John and Joanna were both Irish immigrants. At the age of 14 she was working in a tobacco factory, where everybody called her "Maggie." In 1885, eighteen-year-old Margaret followed her sister Mary Ann and her new husband to Leadville, Colorado. She got a job in the local Mercantile selling carpets and drapes. In the early summer of 1886, she met James Joseph "J.J." Brown. Brown was a miner whose parents had also emigrated from Ireland. They married in September and moved into Johnny's cabin. Margaret had always planned to marry for money, but in the end she chose love.

> *"I wanted a rich man, but I loved Jim Brown. I thought about how I wanted comfort for my father and how I had determined to stay single until a man presented himself who could give to the tired old man the things I longed for him. Jim was as poor as we were, and had no better chance in life. I struggled hard with myself in those days. I loved Jim, but he was poor. Finally, I decided that I'd be better off with a poor man whom I loved than with a wealthy one whose money had attracted me. So I married Jim Brown."*

In 1893 the repeal of the Sherman Silver Act nearly ruined Leadville, but the man they called *"Leadville Johnny"* had a plan.

Jim devised a method to shore up the silver mine and turn it into a gold mine. As a result, he was awarded 12,500 shares of stock and a seat on the board of his employers. The Browns' were rich and Margaret was off and running. She became active in numerous charities, philanthropic causes, attended Carnegie-Mellon, was a pioneer in the feminist movement and although she couldn't vote, ran for The United States Senate. Along with her daughter Helen she traveled extensively and in the spring of 1912 was staying with the Astor party in Cairo, Egypt, when word came that her grandchild was ill. She decided to leave for New York, and [like the Astors] booked passage on the earliest ship home. Helen, a student at the Sorbonne, chose to stay in London. As a result of her last minute change of plans, almost no one, including her family, knew she was sailing home on the *Titanic*. She would come aboard a relatively unknown First Class passenger, but destiny had plans for "Maggie" Brown.

Quigg Baxter joined *Titanic* in Cherbourg, along with [officially] his mother and sister. Unofficially, there was a fourth member of their party. Baxter was the son of James Baxter, a Montreal banker. Dad was credited with opening the first shopping mall in the city, and spending five years in jail for embezzling $40,000 from a bank.

In spite of his father's malfeasance, the family led a very comfortable life. Quigg was a world-class athlete, until a hockey accident cost him the sight in one eye and ended his career. In 1911, he dropped out of his first year at McGill University to accompany his mother and sister to Europe. While in Brussels, he met and fell in love with a 24-year-old cabaret singer, Bertha Mayné. He was determined to bring her back to Montreal with him, and she was booked into a stateroom of her own, [C90] under an assumed name, Madame DeVilliers.

Quigg Baxter, First Class passenger, *RMS Titanic*

Helen Candee, First Class passenger, *RMS Titanic*

The ship also boasted its own "writer's group" composed of seven First Class passengers, all of them writers in one respect or another, [two were published]. After meeting aboard the ship, they gathered every evening of the voyage in one of *Titanic's cafés* to discuss not only their own work, but also various works of literature popular in both Europe and America. And who was at the center of the group? Does the phrase *"cherchez la femme"* mean anything to you?

One of the most popular travelers in First Class was a woman from Connecticut named Helen Churchill Candee [*née* Hungerford]. Edwardian correctness demanded that if a woman was traveling alone, available gentlemen should offer to be her *"protector."* Mrs. Candee was a fifty-two-year-old divorcee of considerable beauty, wit and charm, thus the line of swains was quite long. Archibald Gracie, [he of the Civil War snoozer, *The Truth about Chickamauga*] the self-anointed leader of the group and a member in good standing, dubbed their circle *"our coterie."* Their stated goal was to insure that the comely Mrs. Candee was never left alone. The gentleman who vied for her favor ranged in age from twenty-eight all the way to fifty-eight. For her part, she was flattered by all the attention. In the morning, Helen enjoyed getting a little sun and fresh air on the Promenade Deck. For this, she required two deckchairs, one for herself and another for any admirer who just might happen by.

A woman ahead of her time, Mrs. Candee was a strong believer in the rights of women. In 1900, Helen authored the best selling book, *"How Women May Earn a Living."* In all, she wrote eight books, and became a member of the Washington D.C. social scene. She counted among her friend's political opposites, First Lady Helen Herron Taft, and William Jennings Bryan. While traveling in Europe in the spring of 1912, she received a telegram from her daughter.

Archibald Butt, First Class passenger, *RMS Titanic*

President William Howard Taft with Major Archibald Butt

Her son Harold had been been injured in an auto accident. She cancelled the rest of her tour and headed for home on the *Titanic*.

Archibald Willingham DeGraffenreid Clarendon Butt was born in September 1865 in Augusta, Georgia. The Army was in his blood. His grandfather, Archibald Butt, served in the American Revolutionary War, and his great-grandfather, Josiah Butt, was a Lieutenant Colonel in the Continental Army during the same conflict. He was the nephew of General William R. Boggs of the Confederate States Army. In college, he became interested in journalism and during his senior year, was named editor of the school's newspaper. In 1993 he moved to Washington, D.C. and eventually became the national affairs editor for *The Atlanta Constitution.*

Butt's own military career began when he volunteered during the Spanish-American War, and served with the Quartermaster Corps. He was recalled to Washington in March 1908. Within a month he became the military aide to President Theodore Roosevelt. When William Howard Taft became President in March 1909, he asked Butt to stay on. By 1912, Taft's first term was coming to an end. Roosevelt, who had fallen out with Taft, was known to be considering a run for President against him. Close to both men and fiercely loyal to each, Butt began to suffer from depression and exhaustion. In February, Taft ordered him to go on vacation. He left on a six-week tour of Europe and booked a cabin [B38] home on the *RMS Titanic.*

The couple in staterooms A16/20 had inexplicably, come aboard as "Mr. & Mrs. Morgan." This little ruse, however, fooled no one. Sir Cosmo Edmund Duff-Gordon and his wife Lucy were known by one and all. Duff-Gordon was the 5th Baronet of Halkin and a wealthy Scottish landowner and sportsman.

Sir Cosmo Duff-Gordon, First Class passenger, *RMS Titanic*

Lady Lucy Duff-Gordon, First Class Passenger, *RMS Titanic*

He was particularly noted for his prowess with a fencing foil, and represented Great Britain at the 1906 Summer Olympics, winning silver in the team *épée* event. In 1900, Duff-Gordon married *"Madame Lucile"* [née Lucy Christiana Sutherland] the famous London fashion designer. The whole affair being slightly risqué, since Lucy, the new Lady Duff-Gordon, was a divorcee and had a sister, Elinor Glyn, "famous" for authoring erotica. Along with them was Lucy's secretary, Laura Francatelli [better known as "Franks"] in her own First Class Cabin [E36]. Mrs. Duff-Gordon was travelling to America on business in connection with the New York branch of her Salon. They boarded *RMS Titanic* in Cherbourg.

The gentleman in cabin F2 brought with him two sons and a secret. Michel Navratil was born in Slovakia. In 1902, he moved to Nice, France and became a tailor. In May of 1907 he married an Italian woman, Marcelle Caretto. They had two sons, Michel and Edmond Roger. By 1912, the business was in trouble and Michel claimed that Marcelle had been having an affair. The couple then separated, the boys going with their mother. The sons went to stay with their father over the Easter weekend, but when Marcelle came to collect them, they were gone. Navratil had decided to take the boys to America.

They boarded the *Titanic* at Southampton, the boys being booked as "Loto" and "Louis." He came aboard as Louis M. Hoffman, the "borrowed" name of his friend Louis Hoffman, who helped him to escape France. He led his fellow passengers to believe that "Mrs. Hoffman" was dead and he rarely allowed the boys out of sight. Uncertain of his future, Navratil wrote to his mother in Hungary. The letter, sent from the ship asked if his sister and her husband could care for the boys, if they couldn't stay in the United States.

Michel Navratil, Second Class passenger, *RMS Titanic*

Michel and Edmond Navratil, Second Class passengers

In Second Class [D56] was Lawrence Beesley. Beesley, a widower, had quit his position as a teacher to go on holiday in America, and visit his brother in Toronto. He enjoyed all the amenities provided by White Star and most of all, the free one courtesy of Mother Nature.

> *"Each night the sun sank right in our eyes along the sea, making an undulating glittering pathway, a golden track charted on the surface of the ocean which our ship followed unswervingly until the sun dipped below the edge of the horizon, and the pathway ran ahead of us faster than we could steam and slipped over the edge of the skyline."*

The crew of the *Titanic* numbered nearly 900, of which only 23 were women. Among them was a stewardess named Violet Jessop. She was the first of nine children, only six of whom survived. Born in Argentina, her family moved to England following the death of her father, where she attended a convent school. After her mother became ill, she left school to look for work. At the age of 22, in spite of a dislike of the sea, she hired on to the *RMS Olympic* to work as a stewardess. Jessop was aboard the *Olympic* when she collided with *HMS Hawke*. With her ship laid-up for repairs, Violet was given the chance to join the newest White Star Liner. She was happy on the *Olympic* and didn't want to join the *Titanic,* but was persuaded by her friends who thought it would be a *"wonderful experience."*

As on all the great liners, the *Titanic* featured live music. The ship's eight-member orchestra [deemed an orchestra rather than a band, and including a pianist, a bassist, three cellists, and three violinists] boarded at Southampton and travelled as second-class passengers. Led by bandmaster Wallace Hartley, they played at teatime, during and after dinner, and at Sunday Services - always in First Class.

Violet Jessup, Stewardess, *RMS Titanic*

Wallace Hartley, Bandmaster, *RMS Titanic*

The Anderssons and friends, Kisa, Sweden

Hartley was born and raised in Colne, England. His father was the choirmaster at their church and he learned to play the violin from a fellow parishioner. While working with the choir, he taught them a new hymn, *"Nearer My God To Thee."* In 1909, he joined the Cunard Line as a musician serving on, among others, the firm's newest ships, *RMS Lusitania* and *RMS Mauretania*. In April 1912, White Star poached Hartley away from Cunard and made him the bandmaster of *RMS Titanic*. He left his fiancée, Maria Robinson, to join the ship in time for her maiden voyage, hoping that working on the new liner would enhance his reputation with both White Star and other firms. [His violin would become the center of a controversy a century later.]

In total, passengers and crew exceeded 2,200. First Class included Major Arthur Godfrey Peuchen, President of Standard Chemical and San Francisco banker, Dr. Washington Dodge. In Second Class could be found Mrs. Ada Ball, moving to Jacksonville, Florida, to live with her sister. Also traveling in Second were Ennis Watson and the rest of the Harland & Wolff Guarantee Group. In Third Class there was Olaus Abelseth, one of a group of six friends headed for Minnesota, and the Sage family, all eleven of them. They too were relocating to Jacksonville, to start a pecan farm. And then there were the Anderssons, mother, father and five children. Anders, Alfrida and their children, Sigrid, Ingeborg, Ebba, Sigvard and Ellis, were on their way to Winnipeg, traveling on one £31 family ticket. Their entire hometown of Kisa, Sweden had turned out to see them off.

By Saturday *RMS Titanic,* her crew of 885 and 1,343 passengers were well out into the North Atlantic. With good weather, Captain Smith had ordered the engineers to increase her speed. In the past 24 hours she had covered 519 nautical miles, for an average of 21.6 knots.

White Star advertisement poster, *RMS Titanic*

Chapter 6
Forewarned

*"I know this isn't scientific,
but this ship's warning me
she's gonna die and take
a lot of people with her."*

Thomas Andrews

> "We desire to direct your attention to the Company's Regulations for the safe and efficient navigation of its vessels, and also to impress upon you in the most forcible manner the paramount and vital importance of exercising the utmost caution in the navigation of the ships, and that the safety of the passengers and crew weighs with us above and before all other considerations. You are to dismiss all idea of competitive passages with other vessels, and to concentrate your attention upon a cautious, prudent and ever watchful system of navigation which shall lose time or suffer any other temporary inconvenience rather than incur the slightest risk which can be avoided."
>
> Instructions to Commanders
White Star Line

By 1912 the wireless, the brainchild of Guglielmo Marconi was fifteen years old. He founded The Wireless Telegraph & Signal Company [*The Marconi Company*] and started equipping ocean liners with sets. His invention came of age in 1909, when the White Star Liner *Republic*, one day out of New York and bound for England, collided with the Italian cargo ship *Florida,* in the icy waters off Nantucket. The *Florida* had penetrated her hull and the liner was sinking. The collision killed two of her passengers and damaged the wireless. The operator on duty, Jack Binns quickly made repairs and began to transmit, "*MKC*" [call sign for the *Republic*] "*CQD*" [code for distress]. Binns had just sent the first call for help ever made by wireless. Although his signal was weak and he worked from batteries alone, he reached the Siasconset wireless station on Nantucket. The ship's sole operator, he stayed at his post for the next thirty-six hours, sending signal after signal from his frigid, flooded cabin.

Guglielmo Marconi, inventor of the wireless

The next day another White Star liner, the *Baltic*, picked up the signals and came to the rescue. Over fifteen hundred passengers and crewmen were saved. Binns was a hero, Marconi a genius, the doubters had been silenced, and the wireless was here to stay.

By the 14th of April, the *Titanic's* state-of-the-art set had been receiving ice warnings for four days. Those warnings included six on the 11th of April from ships that were either stopped in or passing through heavy ice fields. The wireless operators dutifully received, cataloged and passed the messages to the bridge. Meanwhile, many of her passengers had discovered the usefulness of Marconi's invention. Personal traffic was flowing in a steady stream. There successes notwithstanding, wireless operations in 1912 were still less than ideal. Some sets worked for hundreds of miles, other far less. The system [as we will see] was failure prone. Messages were often garbled and would have to be repeated. All the sets shared the same frequency, thus turning the North Atlantic into one giant party line. Worst of all, most of the smaller vessels carried only one operator, thus were unable to keep a 24-hour watch.

The dangers of ice in mid-ocean in April were hardly a secret. Ships traveling between the Continent and the New World followed the shortest possible path known to all sailors as *The Great Circle Route*. From August 24 until January 14 Liners would use the shorter Northern Track. The rest of the year, ship travel was via the Southern Track, 110 miles longer but further from the ice. In theory vessels would be able to skirt around the area near the Grand Banks, commonly know as *Iceberg Alley*. Titanic was on the Southern Track. She followed *The Great Circle Route* from Fastnet Light across the Atlantic to a point called "the corner" at 42' N, 47' W.

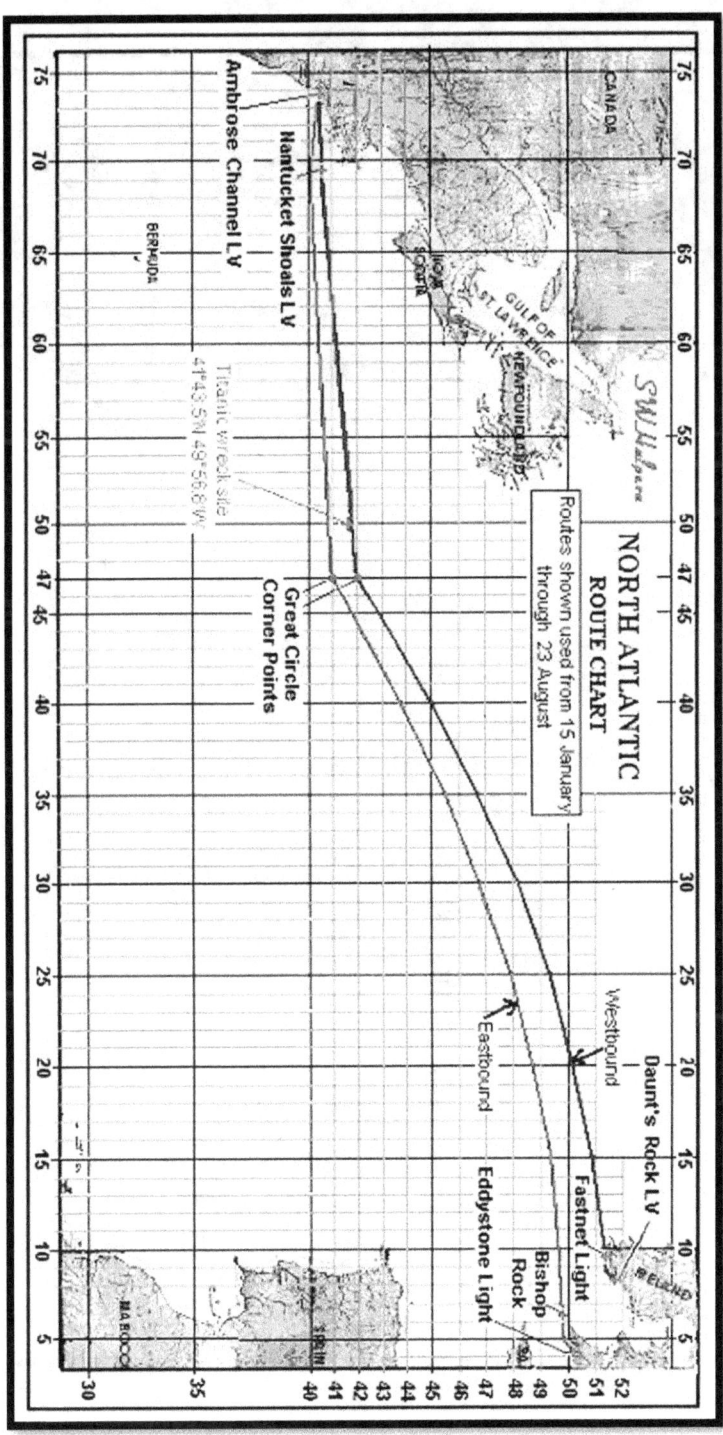

Southern Great Circle Routes, North Atlantic

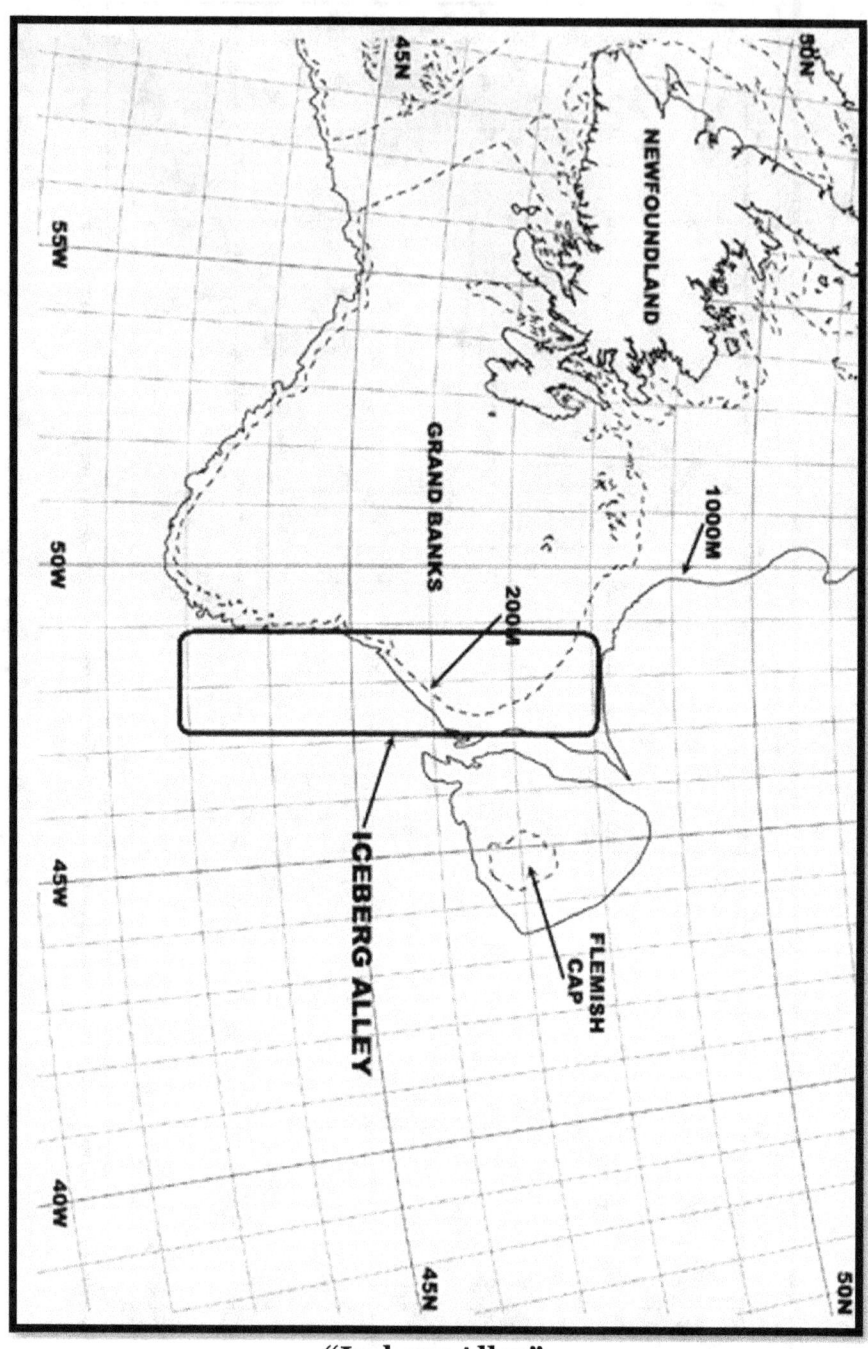

"Iceberg Alley"

This was the turning point for westbound steamers. From there she would have a straight shot to a point just south of the Nantucket Lightship. Twelve hours later the liner would reach *"The Narrows,"* the tidal straight that is entrance to New York Harbor.

Five more "heavy" ice warnings arrived on Friday the 12th. One came at 7:00am from the French Line's *La Touraine*. At the time she was approximately sixty hours ahead of the White Star liner. Her message, specifically addressed to the *Titanic,* was passed on to Captain Smith and posted on the bulletin board in the chart room for all of the officers to see. Phillips and Bride entered it in the logbook and sent an acknowledgement back to the French ship.

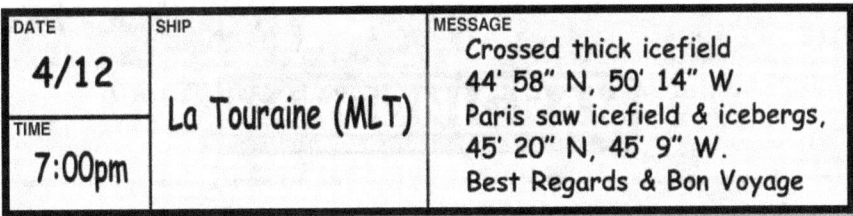

DATE	SHIP	MESSAGE
4/12	La Touraine (MLT)	Crossed thick icefield 44' 58" N, 50' 14" W. Paris saw icefield & icebergs, 45' 20" N, 45' 9" W. Best Regards & Bon Voyage
TIME 7:00pm		

In the foreword of the book I mentioned myths and conjecture. So, now we take up the story of the *Rappahannock*. The *Rappahannock* was a small steamer en route from Halifax to England. That much is true. The story says that at around 10:30pm she passed close enough to signal the *Titanic* with her Morse Lamp [she had no wireless]. The signal: *"we have just passed out of an ice field and we are damaged."* *Titanic* acknowledged receiving the signal and continued on course. [There is no testimony at either Inquiry about such a *"signal."*] There are two versions of this story. One says it happened on the evening of the 13th and the other says it was the 14th. Historians have gone round and round on this one, even though neither seems at all plausible. One is most unlikely and the other is idiotic. 10:30pm on the 14th was one hour and ten minutes before she struck the berg.

William Murdoch, First Officer, *RMS Titanic*

Rappahannock

Does anyone believe that an officer of Murdoch's experience would be foolish enough to go barreling along at over 22 knots if there was an ice field just beyond the horizon? As for the 13th, depending on the two ships exact location, they would be somewhere in the neighborhood of 500 miles apart. As any attorney would stell you, *"Just stipulate that this event did not take place."* I so stipulate.

Shortly after 11pm on Saturday, April 13th, *Titanic's* Marconi system unexpectedly stopped working. In spite of the fact that he had minimal training in repairing the set, Phillips took the wireless apart. He worked through the night before finding the burnt wire that grounded the system. It was 5am before the set running again. Phillips and Bride were now six hours behind schedule. They were sitting on a pile of personal messages, but they would have to wait. Ice warnings were starting to arrive. At 9:00am the *Titanic* received one from [Captain] Barr on the White Star's *Caronia*. The message was delivered to the bridge, and at the Captain's request, posted for his officers to take note. The wireless was working again, for now.

DATE	SHIP	MESSAGE
4/14	Caronia (MSF)	West-bound steamers report bergs, growlers, and field ice on 42' N, from 49' to 52' W. April 12th, Compliments, Barr
TIME 9:00am		

After receiving the message, the Captain set out on his Sunday morning rituals. First would come an ancient tradition: The ship's inspection. With his department heads in tow, Smith covered every deck through every part of the ship. Normally, the next stop would be lifeboat drill. Except that on this particular Sunday, it was cancelled. This came as a surprise to experienced travelers, as was the fact that no reason was given why.

The next scheduled event was church services. At 11am there was a Church of England service that was presided over by Captain Smith in First Class, and Father Thomas Byles conducted a Catholic mass in the Second Class lounge, and then one for the Third Class passengers. At the conclusion of the service, the Captain returned to the bridge for the next scheduled event, the noon sighting. As he walked, Phillips and Bride were taking in another ice message, this one from the Dutch ship *Noordam*, relayed by Cunard's *Caronia*.

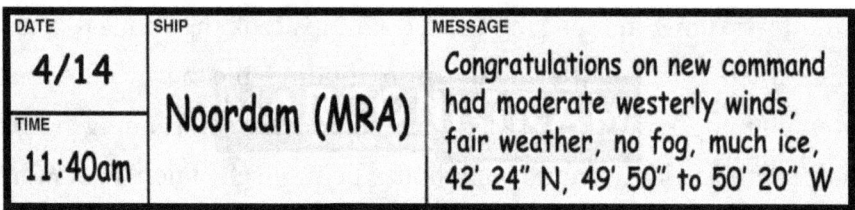

DATE	SHIP	MESSAGE
4/14	Noordam (MRA)	Congratulations on new command had moderate westerly winds, fair weather, no fog, much ice, 42' 24" N, 49' 50" to 50' 20" W
TIME		
11:40am		

Each day Smith and his officers would gather on the portside bridge wing. There, with sextants in hand, each officer would take a noon sun sighting. With their computations they could, with a fair degree of accuracy, work out the ship's position. The results were also used for something else. By comparing yesterday's noon position with today's they could calculate the distance covered in the last 24 hours. The number was posted for all to see, and was a matter of great importance to those high rollers in First Class who had money riding on the results - and the results were impressive. Over the last 24 hours the *Titanic* had covered 546 nautical miles at an average speed of 22.75 knots [her design speed was 22.5 knots]. The big liner was exceeding expectations and there was, it seemed, more to come. The engineers had lit off two more boilers and a rumor was circulating that Captain Smith intended to fire up the remaining three on Monday morning to see what she could do. Twenty-three or even twenty-four knots [wishful thinking] could be within reach.

The speed of a ship, any ship, is ultimately the Captain's responsibility. The story persists that Smith was after a record run. This, of course, was utter nonsense. The *Titanic* couldn't run down the *"Maurey"* with every boiler lit and the safety valves thrown overboard. [She had the looks, but the *Mauretania* had the legs.] Any chance of capturing the *Blue Ribband* was lost in the design. Much would be written [mostly in Hearst newspapers] that Ismay had ordered or at the very least, urged a fast crossing. This is most unlikely. Even though the Managing Director enjoyed using his influence, none of this would have mattered to Smith. White Star Vice President P.A.S. Franklin had floated a concept for a faster crossing. In his plan their ships would arrive in New York on Tuesday night rather than the planned Wednesday morning. To Ismay, this proposition was very much a non-starter.

> *"I at once admit that docking on Tuesday evening would help you in turning the ship round, and give those on board a better chance of getting the ship in good shape for the Saturday sailing, and further, that if we could make it a practice to do this, it would please the passengers, but as I have repeatedly stated, I feel very strongly that passengers would be far more satisfied to know, when they left here, that they would not land until Wednesday morning, rather than be in a state of uncertainty in regard to this for the whole of the trip. I do not think you can have ever experienced the miseries of a night landing in New York; had you done so, I think your views might be altered."*

The *Olympic*-class liners, in spite of all their excesses, had been designed to be efficient and economical to operate. If the Line wanted a faster time they would have opted for Parson's Turbines.

Philip Albright Small Franklin, White Star Line

René Jacques Lévy and daughters

They would have junked their now inferior propulsion system. To Ismay the added expense in coal and the wear and tear on the equipment would be cost prohibitive. As for Smith, he was doing what every other Captain on the *"Atlantic Ferry"* would do. In good weather, run at best possible cruising speed.

Chemist René Jacques Lévy was returning from a family funeral in Paris. He was originally scheduled to sail back to Canada on the maiden voyage of the *France* on April 20, but switched passage when he learned he could get back to his family in Montreal ten days earlier aboard the *Titanic*. After lunch, Lévy was on deck together with Marie Jerwan and cabin-mate, Jean-Noël Malachard [of Pathe Films]. Lévy pointed at one of the lifeboats and said:

> *"I'm sure, if they lower these boats, the falls will be too short. Of course I would prefer to go down with the ship rather than sitting in one of these boats."*

At 1:42pm the *Titanic* received an ice warning from the *Baltic* and the message was delivered to Captain Smith, who discussed it with the Managing Director. This was the message the Wideners saw change hands. Ismay then put the ice warning in his pocket.

Marian Thayer finally got Emily Ryerson on-deck for the first time during the voyage. Ryerson was mourning the loss of her son. The two friends walked for nearly an hour before settling into deck chairs to enjoy the sunset. Ismay approached them. He sat down, and after asking if the ladies were comfortable and enjoying the trip, he explained to them about the possibility of meeting icebergs in the area. He showed them the ice warning from the *Baltic* that Smith had passed to him. Later in the afternoon, the Captain asked for the return of the warning, and he posted it in the chart room at 7:15pm.

DATE 4/14	SHIP	MESSAGE
TIME 1:42pm	Baltic (MBC)	Athenia reports passing icebergs and large quantities of field ice today in latitude 41' 51" N, longitude 49" 52' W. Wish you and Titanic all sucess.

Three minutes later at 1:45pm the *Titanic* received an ice warning from the *Amerika*. This was a private message to the Unites States Hydrographic Office in Washington, D.C. The wireless set on the German liner wasn't powerful enough, and their operator asked Phillips to pass it on. He did, and made a copy for the ship, but inexplicably, did not send it on to the bridge.

DATE 4/14	SHIP	MESSAGE
TIME 1:45pm	Amerika (DDR)	Amerika passed two large icebergs in 41' 27" N, 50' 8" W on April 14.

There would be no ice warning messages between 4pm and 7pm. In fact, no messages of any kind, the wireless had failed again. Three precious hours were lost while Philips made repairs. Had all the messages received up to this point been sent along, or retrieved from the owner's pocket, and then plotted on the chart, it would have shown a huge field of ice directly in the ship's path. No more proof would have been necessary, yet more was coming.

Down on D-Deck, First Class passengers were enjoying another sumptuous meal in the dining room. For Lucy Duff-Gordon, it had been a splendid four days on the *Titanic*.

> "The first days of the crossing were uneventful. Like everyone else, I was entranced by the beauty of the liner.

I had never dreamed of sailing in such luxury ... my pretty little cabin, with its electric heater and pink curtains, delighted me, so that it was a pleasure to go to bed. Everything about this lovely ship reassured me."

"*I remember that last meal on Titanic very well. We had a big vase of beautiful daffodils on the table, which were as fresh as if they had just been picked. Everyone was very gay, and at a neighboring table people were making bets on the probable time of this record-breaking run. Various opinions were put forward, but none dreamed that Titanic would make her harbor that night.*"

The Leyland Line ship *Californian* had sent an ice message to the *Antillian*, which had been overheard and picked up by Bride. The signal went to the bridge. There was no mistaking it's meaning, the *Titanic* was two hours away from the ice. The Captain was not on the bridge. He was the guest of the Wideners' at a dinner party in his honor, in the *á la carte* Restaurant. Smith never saw the ice message.

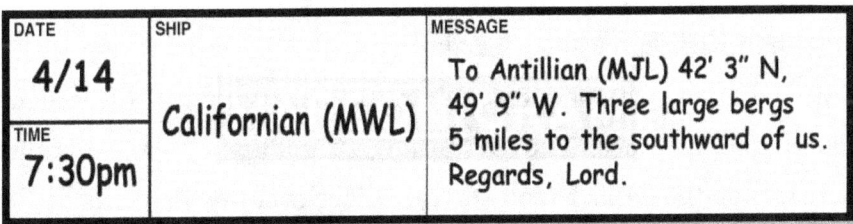

By 9:00pm the passengers were starting to adjourn to their cabins. The temperature had dropped to below 40 degrees, and the combination of the cold and the 25 mph wind generated by *Titanic's* speed had emptied the promenade deck. The water temperature, carefully calculated by dropping a thermometer into a bucket of seawater, had fallen four degrees in the last hour.

On the bridge, Lightoller had the watch. Jim Hutchinson, the ship's carpenter was summoned and instructed to keep an eye on the fresh water, lest it freeze. The dining rooms were deserted, except for the stewards preparing for tomorrow's breakfast. In the *á la Carte* Restaurant the Widener's party was winding up, with the gentlemen heading for the smoking room, their ladies off to the reading room and the Captain back to the bridge. Smith ordered a slight course change, one that moved *Titanic's* track ten miles further to the south.

In the wireless room, Phillips was busy at the key. He had picked up the station at Cape Race and was plowing through the pile of personal messages that had accumulated while the set was down.

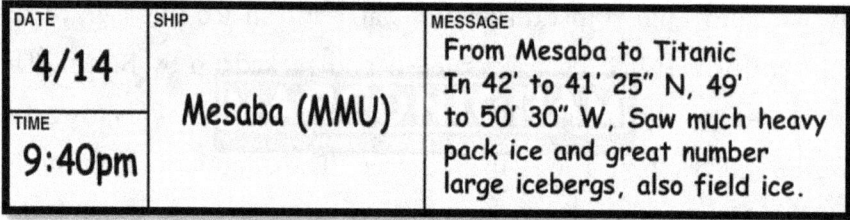

This message, the most serious warning yet, should have gone straight to the bridge. The exhausted Bride was asleep in his bunk and Phillips didn't want to wake him, and he didn't want to go himself because of the personal traffic he was getting from and sending to Cape Race. The ice warning wound up in the pile of other messages on the desk. No one on the bridge ever saw it.

Phillips and Bride worked for *The Marconi Company*, not the White Star Line. Yes, they wore the uniform and yes, they followed orders from the Captain and the officers, but they were on Marconi's payroll. Every personal message sent or received meant money to the Company. Phillips was living the *Parable of the Unjust Steward*, trying to serve two masters.

At 9:20pm Captain Smith retired for the night, leaving orders with Lightoller to awake him *"if it becomes at all doubtful."* At 10pm, the watch changed. The next four hours belonged to First Officer Murdoch, and he took note as the temperature continued to drop. There was no moon, a pitch-black cloudless sky and an ocean that was completely calm. After the fact, crewmen would say that they had never seen the North Atlantic so flat. All over the ship passengers were turning in, while down in the boiler rooms *"the black gang"* kept the furnaces stoked with coal. No matter what speed was called for on the bridge, it was these men, doing their backbreaking work, which made it possible. In the wireless room a tired and exasperated operator Phillips continued his labors.

The *SS Californian* sailed from London on April 5, 1912. The Leyland Liner was inbound to Boston. On this trip it was cargo only, she was not carrying any passengers. During the evening of April 14th, Captain Stanley Lord was on the bridge when the ship encountered an ice field. With no travelers to inconvenience, nor a time-sensitive Royal Mail contract, Captain Lord did the prudent thing. There was no reason to try to feel his way through the ice in the dark. He stopped his ship and shut it down until morning. While still on the bridge, he saw a ship's lights approaching. Lord went to the wireless room to find out if operator Cyril Evans knew of any ships in the area. Evans met him on the way and informed him that there was just one: *"only the Titanic."* Lord instructed him to inform her that the *Californian* was now stopped and surrounded by ice.

The Wireless had become the newest toy for First Class passengers with too much money and too much time. The *"private stuff,"* as the operators would call it, came in a never-ending torrent. Anything and everything now required a telegram.

SS Californian, Leyland Line

Captain Stanley Lord, Commander, *SS Californian*

"Buy 10,000 shares, U.S. Steel at the market price!" "Have my private railroad car ready when I arrive!" "Warmest regards on your upcoming nuptials." All of the messages arrived via pneumatic tube from the Purser's office. Still the somewhat taciturn First Operator trudged on. As Evans sat down at his key and started to send the ice message, *RMS Titanic* was less than 20 miles away.

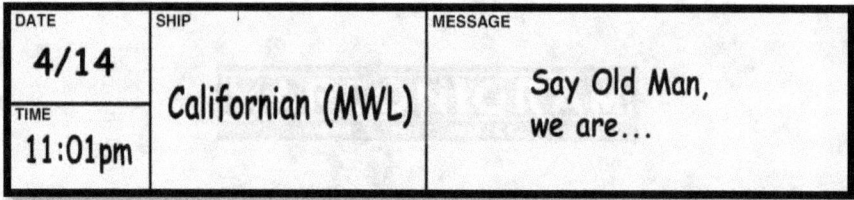

That was as far as Evans got before he was jumped on by an irate Phillips for blasting his ears off from short range and for interrupting his Cape Race traffic without asking permission.

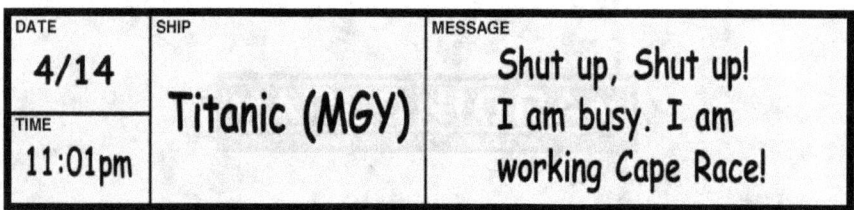

It had been a long day for Evans as well, and the last thing he needed was a rebuke from Phillips. He listened for a while longer and heard *Titanic* raise Cape Race again. At 11:30pm he was finished for the day and turned off the set before going to bed. Both the *Californian* and her wireless were shut down for the night. There was no one to relieve Evans; he was the ship's sole operator.

In the First Class Dining Salon Thomas Andrews, whose slavish devotion to his beautiful but demanding mistress seemingly never ended, now finally appeared satisfied. At dinner he remarked that she was *"as nearly perfect as human brains can make her."*

Chapter 7
A Bump in the Night

*"They that sow the wind,
shall reap the whirlwind"*

Hosea 8:7

At 11:00pm Railroad Executive Charles Hays was sitting in the First Class Smoking Room along with Colonel Archibald Gracie and Captain Edward Crosby. Cigars and brandy mixed with lively conversation. Tonight's subject was the technological advance in transportation. At one point, Hays conceded that while *Titanic* was a *"superlative vessel,"* he expressed concern that *"the trend to playing fast and loose with larger and larger ships will end in tragedy."*

At 11:30pm *Titanic* lookout Frederick Fleet scanned the horizon from his perch in the crow's nest, 100 feet above the surface of the North Atlantic. Tomorrow he and his mate, Reginald Lee, should see the fishermen off the *Grand Banks* and cod boats out of Gloucester, but now they were in the *"alley"* and on the watch for something else, ice. [The bridge had just told them to keep a sharp watch for *"growlers."*] For an hour and a half all they saw was an empty sea. Less than thirty minutes remained before they would be relieved. They could leave their frozen, windswept post in the foremast and go below. Until that time they concentrated on their job, because on this night conditions for spotting were not good. There was no moon to illuminate the ice and the flat, calm sea meant no white waves around the base of a berg. Fleet wished he had his binoculars.

On the bridge Murdoch, along with Fourth Officer Joseph Boxhall and Sixth Officer James Moody, stood the watch. Behind and just above was the wheelhouse, with Quartermaster Robert Hichens at the helm. The air temperature was 30 degrees, while the water was 28 degrees. The bridge chronometer displayed the time: 11:40pm.

From out of the darkness Fleet saw a black mass ahead, blacker than even the night sky. He immediately rang the *"crow's nest"* bell three times [the signal for an object ahead] and telephoned the bridge.

Frederick Fleet, Lookout, *RMS Titanic*

Reginald Lee, Lookout, *RMS Titanic*

Moody picked up the phone and asked, *"What did you see?"* Fleet responded, *"Iceberg right ahead!"* He heard the Sixth Officer say, *"Thank you."* Moody then passed the word to William Murdoch.

The First Officer had a decision to make, and perhaps a second to make it. Like every officer on the transatlantic run, he knew that when faced with a choice of hitting an iceberg head on or sideswiping it, you always took the former. The bow would be crushed, the forward compartments flooded, but the ship would remain afloat. Exposing a ship's flank to a berg was just asking for trouble. Still, if there is a chance to avoid the ice completely, you took it. As soon as the berg was visible from the bridge, Murdoch made his decision. *"Hard a-starboard!"* he bellowed to Hichens, who immediately put the helm hard over. The First Officer then stepped to the engine telegraph and rang up full speed astern on the wing engines. [Remember, the center prop could not be reversed.] After what must have seemed like an eternity, the ship began to respond, slowly turning to port. The lookouts braced for the worst, a collision with the berg. Still the *Titanic* turned left as the berg came closer. At the last second her stem seemed to be in the open. The bow was in the clear but not the rest of the ship. *"Hard-a-port!"* should have been the next order, but by now they were on top of the berg. As they passed, the lookouts head a faint scraping sound as chunks of ice hit the decks along the starboard side. It was then that Fleet and Lee realized why the berg was so hard to spot. It was a *"blue berg,"* meaning that the iceberg had flipped in the last few hours and was showing her seawater stained side. On the bridge Murdoch activated the watertight doors, thereby "sealing" all sixteen compartments.

What you felt and heard was a function of where you were on the ship. If you were in a cabin on the starboard side you heard a loud

Charles Victor Groves, Third Officer, *SS Californian*

Joseph Groves Boxhall, Fourth Officer, *RMS Titanic*

grinding, tearing sound. The lower you were on the ship the louder it got. If you were James McGough in E25 and left your porthole open, you suddenly had a cabin full of ice. Some felt the ship had run over something, and a few even believed she had run aground on Newfoundland. [No small accomplishment in 12,000 feet of water.] On the port side some felt nothing, some would describe a scrapping sound, while others felt a shudder going through the ship.

That shudder was most definitely felt in the First Class Smoking Room, where the last members of the Widener's dinner party for the Captain had a card game going. Archibald Butt, Harry Widener, William Carter and Clarence Moore were all that remained. Moore, a successful broker and landowner, had been in England shopping for foxhounds. The vibration stopped the game in its tracks as the four men jumped to their feet and ran out onto the promenade deck. They reached the rail in time to hear some one shout, *"We've struck an iceberg, there it is!"* A few on the deck got a fleeting look before it melted back into the darkness. A moment later they returned to the warmth of the smoking room and their cards.

Newlyweds Dick and Helen Bishop felt it and left their cabin on B-Deck to investigate. There they met Steward Alfred Crawford. Crawford had seen a large black object float by and told the couple to, *"go back downstairs, there is nothing to be afraid of. We have only struck a little piece of ice and passed it."*

Seaman Frederick Clench was asleep in his bunk and awoke to "*a crunching and jarring sound.*" He jumped up, went up to the well deck and "*saw a lot of ice.*" Right behind him was seaman Frank Osman. The noise of the collision had brought him up from the seaman's mess on C-Deck.

Broadway [and now film] star Dorothy Gibson's bridge game had finally broken up and she returned to her cabin. A moment later she noticed *"a long drawn, sickening crunch."* In cabin D20 the impact awoke Mrs. Martha Stephenson and triggered a flashback to the morning of April 18, 1906, The San Francisco Earthquake. It quickly passed when she remembered that this jolt was nothing compared to the tremors she felt in California.

J.J. Astor seemed calm and collected, as he returned to his suite, following a trip to the deck to investigate the situation. He explained to his young bride that the ship had struck some ice, but it didn't look serious. He was quite relaxed, and Mrs. Astor was not at all alarmed. To the occupant of Promenade Suite B52/54/56, it felt like a solid jar. Joseph Bruce Ismay awoke with a start. He was certain that his brand new liner had struck something, but had no idea what it was.

For the men in Boiler Room No. 6, it was far more than a jar. They heard the warning bell and saw the light flash red above the watertight door. Next came an ear-splitting crash, as the whole starboard side caved in. Suddenly water was surging into the room. There was no time to draw the fires. A few men managed to charge up the escape ladders, the rest leaped through the rapidly descending watertight door just before it slammed shut. They found water gushing into Boiler Room No. 5 as well. The doors were rigged to close simultaneously, leaving bewildered engineers and fireman further aft to wonder what was going on.

It was also far more than a jar to the Third Class passengers, starboard side forward on E and F-Deck. It wasn't a jolt; it was a tremendous crash, strong enough to toss men out of their bunks and onto the floor. On E-Deck there was a wide alleyway that ran the length of the ship, known to the crewmen as *"Scotland Road."* [Named for a Liverpool Street.] Suddenly *"the Road"* was teeming with Third Class passengers moving aft, away from the crash, clutching their children and whatever belongings were within reach.

In the First Class dining room on D-Deck, the stewards were taking it easy. The last passengers had long since gone. They felt a shudder, a vibration strong enough to rattle the glassware on the tables. James Johnson believed he had felt this before. He had served on the *Olympic* when she dropped a propeller blade. His mates were of a like mind, and one of them, anticipating a return to Harland & Wolff [and the *"delights"* of the port] declared, *"Another Belfast trip."*

There would be no trip back to Queen's Island. The scraping, grinding sound was her death knell. A ship designed to sail for 30 years now had a life expectancy of less than three hours. There was no great gash in her hull. This was more insidious. Popped rivets and bent plates resulting in open seams had created slivers along her side, the equivalent of ten square feet of inrushing water, more than enough to do the job. Call it *"death by a thousand cuts."* The attempt at an evasive maneuver had doomed her. She was now slowly, but inevitably, losing her battle with buoyancy.

A moment later the Captain rushed onto the bridge. He had heard the grinding and scrapping in his cabin. He approached his First Officer. *"Mr. Murdoch, what was that?"* *"An iceberg, sir. I hard-a-starboarded and reversed the engines, and I was going to hard-a-port around it, but she was too close. I couldn't do any more."* *"Close the emergency doors,"* snapped Smith. The First Officer replied, *"The doors are already closed."* Following Captain's orders he pulled on the engine room telegraphs, turning them to "Stop."

[He might as well have rung up "Finished with Engines," since the propellers would never turn again.]

Across the water, the *Californian* was now snugged down for the night. At 11:10pm the watch officer, Lt. Charles Victor Groves, saw the lights of a ship, coming from the east on her starboard side. As the ship grew closer he saw more lights, a blaze of lights, from her deck and sides. This was no tramp steamer, but rather a large express liner. After watching the ship pull even with the *Californian,* Groves went for Captain Lord, who was sleeping on the settee in the chart room. Lord instructed him to try contacting her with a Morse Lamp. As he prepared to send a message, he observed the liner slow, then stop and put out most of its lights.

Groves was a veteran officer and knew that on some secondary runs [passages to India] ships routinely put out their lights in the evening to encourage passengers to go to bed. This, however, was the *"Atlantic Ferry."* No express liner would stop and turn off its lights without a very good reason. It never occurred to him that the lights were still on and that the ship had turned ninety degrees so he couldn't see them. Nor did it occur to him to ask the Captain to wake up Evans, the wireless operator, and have him try to contact the liner.

On the *Titanic*, the sounds and sensations of the collision did not seem to bother many people. What was bothersome was the absence of sound and sensation. Suddenly there was no hum from the turbine, nor gentle vibration from the engines. It was this that made passengers summon stewards, dispatch bartenders and stop seamen. Soon there was a new, unfamiliar sound from deep within the ship. The Liner's engineers were diverting steam from the motors to the pumps. In stateroom C51, Archibald Gracie had gone to bed early.

> "I was awakened in my stateroom at 12 o'clock. The time, 12 o'clock, was noted on my watch, which was on my dresser, which I looked at promptly when I got up. At the same time, almost instantly, I heard the blowing off of steam, and the ship's machinery seemed to stop."

Captain Smith wanted a damage assessment and he wanted it now. He sent Boxhall below. When the Fourth Officer returned he reported that he didn't see any damage. *"E. J."* was not satisfied. He called for the ships carpenter Jim Hutchinson to *"sound the ship"* [check for damage and if possible, make repairs]. As Boxhall ran down the staircase, Hutchinson pushed past him on his way to the bridge, yelling, *"She's making water fast!"* Chief Officer Wilde arrived, and a moment later, Ismay joined them. Smith gave them the news about the berg. The Managing Director asked the Captain if the ship was seriously damaged. *"I'm afraid she is"* came the reply.

The question was, *"just how serious?"* The Captain then called for the one man aboard [arguably the one man in the world] who could give him a definitive answer, Thomas Andrews. Andrews had been in his cabin [A36] going over plans for "improvements" he could make to the *Titanic* [and no doubt, the *Gigantic*]. Upon arrival, he and Smith immediately left for their own tour of the ship. As they walked off the bridge, the ship's commutator showed *Titanic* with a five-degree list to starboard and already two degrees down by the head.

Things seemed as dire as reported. Using passageways normally reserved for the crew, they saw holds flooded, clerks moving mail away from the rising tide and from the gallery, water in the squash court. They returned to Andrews' cabin, only this time in sight of passengers. Neither man's face would betray what he felt inside.

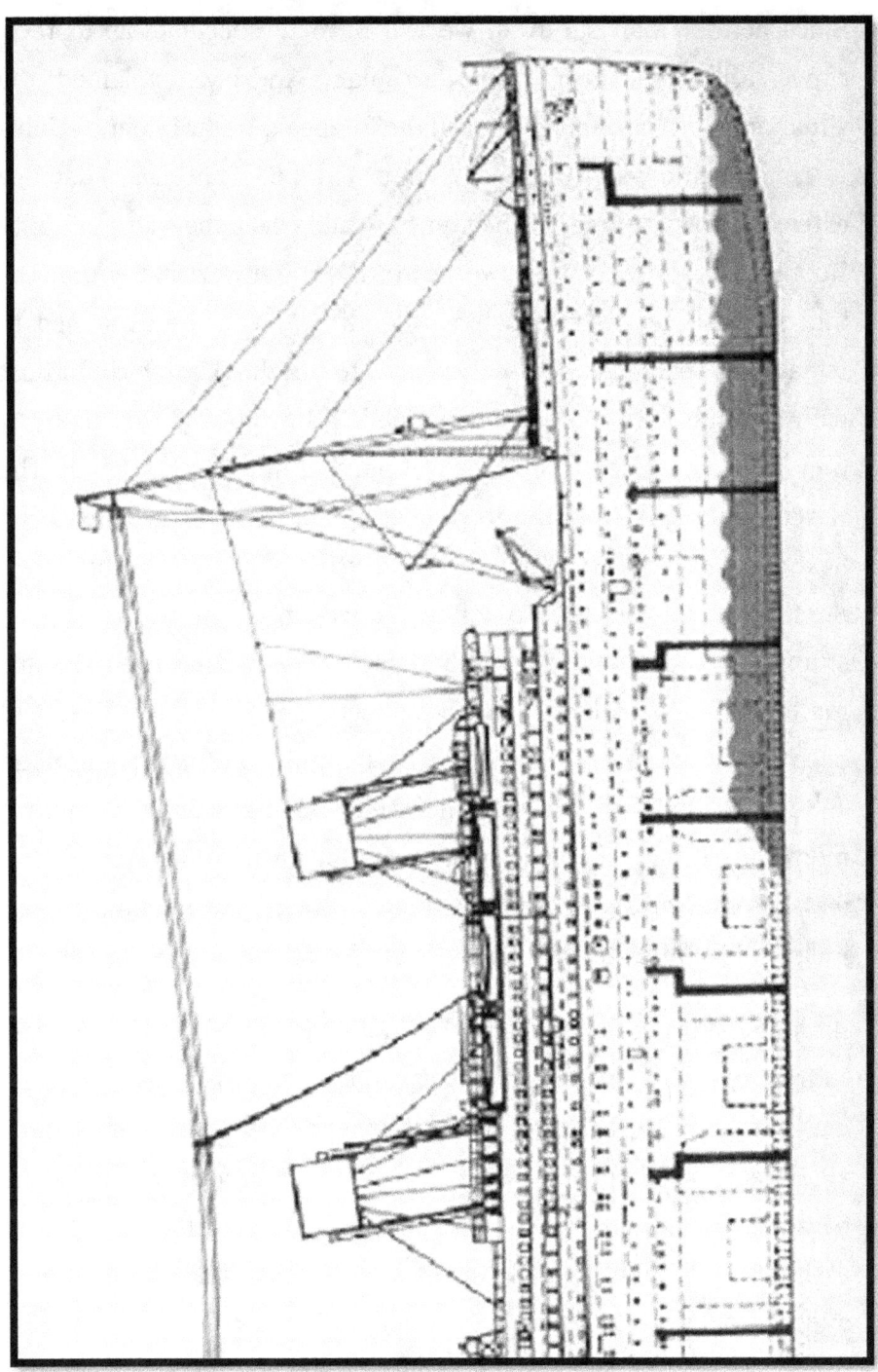

Initial flooding, *RMS Titanic*

Since neither man survived, we will have to speculate as to their conversation. With a side view plan, Andrews explained the situation. She was taking water in the forepeak, water in Cargo Holds 1, 2, and 3, and water in Boiler Rooms 5 and 6. That meant damage extending some 300 feet. The pumps would check the water in Boiler Room No. 5 [only the first two feet of the room were compromised] but the first five were hopelessly flooding. What he explained to Ismay four years ago, he now repeated to Smith. *Titanic* could float with her first four compartments flooded, but not the first five. As they fill, the weight of the water will pull her down by the head. The water would spill over the top into the next compartment and the next and so on. It was metaphysical certitude, she was going down. Smith [his life at sea no longer *"uneventful"*] must have been stunned. Andrews was surely heart-broken, knowing that his masterwork was sinking beneath his feet. The last time a liner had been lost to an iceberg was 1856. The shipping lines had been rolling sevens for sixty-six years. Their luck changed in a matter of moments. It had come up "snake eyes" for the *Titanic*. And now, it was left to Smith and Andrews to pay the price for Ismay's poor judgment. White Star's chickens had indeed come home to roost.

Back on the bridge at 12:05am, the Captain gathered his Lieutenants. In front of Smith stood his trump cards, career British Naval Officers, who had spent their lives at sea preparing for a moment such as this. He ordered Wilde to uncover the boats and Murdoch to muster the passengers. Sixth Officer Moody was sent to get the list of boat assignments, while Boxhall was charged with waking up Lightoller and Third Officer Pitman. Smith himself walked to the wireless cabin. It had been an hour since John Phillips told the *Californian* to "*shut up,*" but now, at last, he was caught up with the *"private stuff."*

Oblivious to the fact that *Titanic* had stopped, the First Operator was not scheduled to be relieved until 2:00am, but Bride was already awake and ready to take over. A moment later the Captain appeared.

> "We've struck an iceberg and I'm having an inspection made to see what it has done to us. You'd better get ready to send out a call for assistance, but don't send it until I tell you."

He left, but returned moments later. *"Send the call for assistance."* Phillips asked, *"The regulation call?"* *"Yes, at once."* The Captain handed Phillips her position and left. The First Operator sat down at the key, took the headphones from Bride and started sending.

DATE	SHIP	MESSAGE
4/15	Titanic (MGY)	CQD! CQD!
TIME		41' 44" N
12:15am		50' 24" W

"CQ" was a general call to all ships. The critical letter was the *"D,"* for distress. Six times Phillips sent the same message. Across the seas the message went out by wireless, around the ship it was passed by a more traditional time-honored method, word of mouth. It got from the Orlop Deck to the crew's quarters in a hurry.

If your berth was forward, you only needed to look into one of the cargo holds to see there was trouble. Trimmer Samuel Hemming and Boatswains Mate Albert Haines wondered why they heard the sound of air escaping from the forepeak tank. It was the sound of air being displaced by the water that was now flooding the tank.

The officer's quarters were located behind the bridge. Second Officer Lightoller had returned to his cabin after making the rounds and at 11:40pm, was just nodding off when he felt a grinding vibration.

Herbert John Pitman, Third Officer, *RMS Titanic*

Charles Herbert Lightoller, Second Officer, *RMS Titanic*

Still in his pajamas, he went on deck where he met Third Officer Pitman who had also felt the vibration. [Pittman would testify that it felt like *"coming to an anchor."*] They concluded that the liner had hit something but could see no sign of trouble. And since there was no evidence of undue alarm on the bridge, they returned to their cabins. Lightoller got back in bed and waited. If they needed him he should be where they would expect to find him. He knew his duty, knew it to the letter. Ten minutes later, Fourth Officer Boxhall burst into his cabin: *"The water is up to F-deck in the Mailroom!"* That was all he needed to hear. Lightoller pulled on a pair of trousers, a sweater and a bridge coat over his pajamas and headed to the bridge. The man the crew called *"Lights,"* [except when he was around] was ready. Walter Lord had it right: *"He was the perfect Second Officer."*

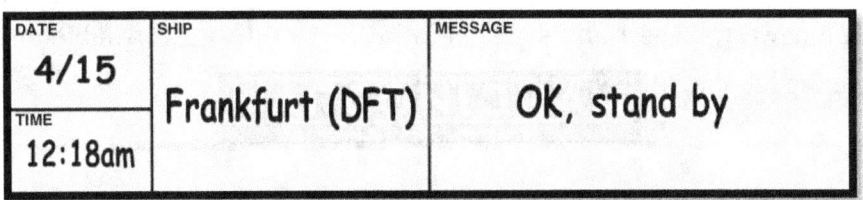

DATE	SHIP	MESSAGE
4/15	Frankfurt (DFT)	OK, stand by
TIME		
12:18am		

The word spread rapidly. Cape Race heard it and relayed it to Halifax. The wireless station at Siasconset on Nantucket Island heard it. Ships in range of the *Titanic* heard it and passed it on to steamers farther away. Soon the whole world would know the news. Guglielmo Marconi had seen to that.

Outbound from New York, on her way to the Mediterranean port of Fiume, and approximately 58 miles from the *Titanic*, was the 13,000 ton Cunard Liner *RMS Carpathia*. The ship's lone wireless operator, Harold Thomas Cottam was [like Evans] finished for the night. He was on the bridge, chatting with the officers on watch, when Phillips started sending *"CQD."* Ten minutes later he was back in his cabin.

Preparing for bed, he was listening to Cape Race trying to raise the *Titanic*. Cottam, as any competent operator would do, copied it down and then relayed the message.

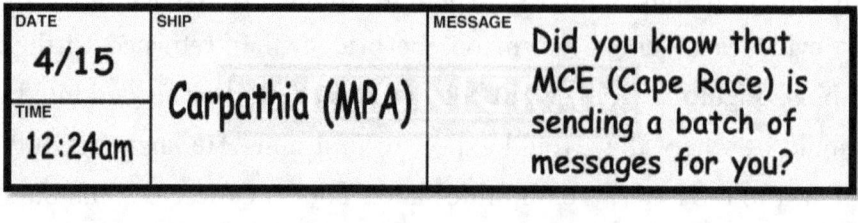

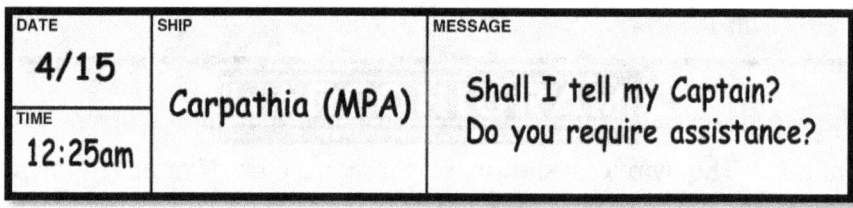

The operator paused, not sure if he believed what he was hearing. The newest, biggest and safest ship in the world was in trouble and calling his ship for help. He replied:

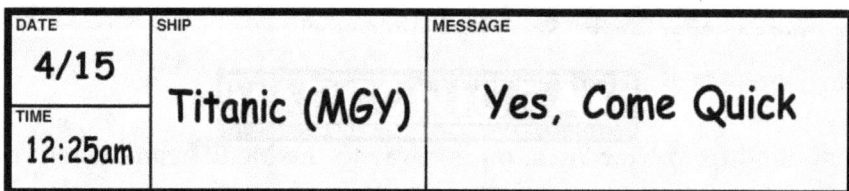

DATE	SHIP	MESSAGE
4/15	Titanic (MGY)	Yes, Come Quick
TIME		
12:25am		

The passengers and crewmen about to enter the lifeboats didn't know it, but they had finally caught a break. The Commander coming to their rescue was the man known throughout the merchant service as "The Electric Spark."

RMS Carpathia, Cunard Line

Harold Cottam, Wireless Operator, RMS Carpathia

Arthur Henry Rostron, Commander, *RMS Carpathia*

Arthur Henry Rostron was born in Astley Bridge, north of Bolton, Lancashire, England. At the age of fifteen, he joined the Merchant Navy Cadet School ship *HMS Conway* as a cadet. A year later he was apprenticed to the Waverley Line in Liverpool, crewing on the Iron clipper ship, *Cedric the Saxon*.

In December 1894 Rostron served on board the steamship *Concord* where he passed the extra master's certificate. He joined the Cunard Line in January 1895 and earned a position as fourth officer on the liner *RMS Umbria*. He also served the Line as a junior officer onboard a half-dozen other ships. As a member of the Royal Naval Reserve, Rostron temporarily left Cunard when ordered to serve with the fleet during a period of international tension as a result of the Russo-Japanese War of 1904–1905.

He returned to the Cunard Line and was appointed First Officer of the *RMS Lusitania* in 1907, but was transferred to the *Bresica* and promoted to ship's Captain the day before the *Lusitania's* maiden voyage. The *Bresica* and his next several ships plied the Mediterranean waters. On January 18, 1912 he was given command of the passenger liner *RMS Carpathia*.

On the *Titanic* the word was passed to the stewards to get the passengers on deck with their lifebelts on. Many got the news directly from Andrews. As they made their way down passageways and around corners, the lifebelts were the only things they shared. A fact not lost on the all-seeing eye of Mrs. Candee.

> *"On every man and every woman's body was tied the sinister emblem of death at sea, and each walked with his life clutching pack to await the common horrors. It was a fancy dress ball in Dante's Hell."*

In B39, Margaritha Frölicher had been seasick since the ship left Queenstown. The *"mal de mer"* finally passed Sunday evening and she was getting some much-needed sleep. She was awakened by the collision. Together with her parents she dressed and went to A-Deck to investigate what happened. Seeing nothing amiss, they went back to bed. Ten minutes later a steward appeared. Now wearing warm clothes and their lifebelts they headed for the Boat Deck. Major Arthur Peuchen [on his fortieth transatlantic voyage] learned from steward James Johnson that the ship had struck an iceberg, but he never believed the ship was going to sink. He grabbed three oranges and a pearl pin from his cabin, but left behind $200,000 in stocks and bonds, jewelry, and the gifts he bought for his children.

In Second Class Albert and Sylvia Caldwell had been on deck until a sailor told them they were in no danger. They returned to bed but were awakened a second time by someone pounding on their doors, a strange voice yelling *"Everyone on deck with your lifebelts!"* The Caldwell's rushed up to the Boat Deck with their year-old son Alden wrapped in a blanket.

In Third Class, Olaus Abelseth somehow managed to assemble his group. This had taken some doing since his charges included men and women and per White Star rules were berthed at opposite ends of the ship. They made their way along *"Scotland Road,"* climbed the Third Class stairs, ultimately winding up on the Poop Deck.

Others in steerage were not so fortunate. If you recall, the Boat Deck was First and Second Class promenade space. Third Class promenade space was nowhere near the boats, and there was no direct access from their section of the ship. In addition, barriers and gates guarded by boatswains and stewards blocked their path.

In the First Class lounge Wallace Hartley's orchestra, all eight strong, had begun playing. Later they would move up to the Boat Deck. Not wanting to alarm the passengers, Hartley chose to keep it light. His initial selections were American Ragtime.

On the bridge Fourth Officer Boxhall had re-calculated *Titanic's* position with a star sight. He brought it to Phillips, who immediately started putting it out over the air.

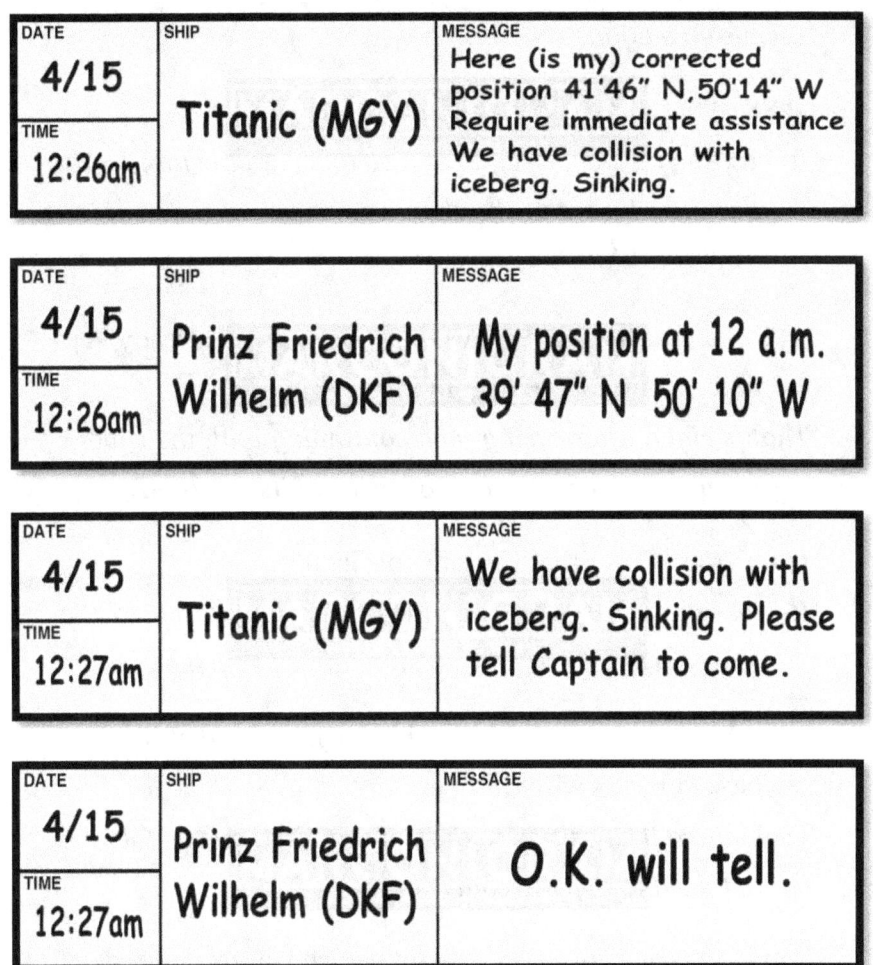

Before he worked out the position, Boxhall saw something on the water. After leaving the Wireless cabin, he returned to the Boat Deck.

Below is a portion of his testimony given at the Board of Trade Wreck Inquiry. The Examiner is Raymond Asquith, Counsel for the Inquiry.

> 15386. Someone reported a light ahead?
> "Yes; I do not know who reported it. There were quite a lot of men on the bridge at the time."
>
> 15387. Did you see the light?
> "Yes, I saw a light."
>
> 15388. What sort of light was it?
> "It was two masthead lights of a steamer. But before I saw this light I went to the chart room and worked out the ship's position."
>
> 15389. Is that the position we have been given already - 41 deg. 46 min. N., 50 deg. 14 min. W?
> "That is right, after seeing men continuing with their work I went to the chart room to work out its position."
>
> 15390. Was it after that you saw this light?
> "It was after that, yes, because I must have been to the Marconi office with the position after I saw the light."
>
> 15391. You took it to the Marconi office in order that it might be sent by the wireless operator?
> "I submitted the position to the Captain first, and he told me to take it to the Marconi room."
>
> 15392. And then you saw this light, which you say, looked like a masthead light?
> "Yes, it was two masthead lights of a steamer."

First Class Grand Staircase, *RMS Titanic*

John "Jack" Thayer, First Class Passenger, *RMS Titanic*

Boxhall was sure of what he saw. Inside the horizon line was a ship. [Maritime Law requires that a steamship show two lights.] A steamer showing masthead lights. Below decks in First and Second Classes stewards urged passengers on. They couldn't order a traveler to put on a lifebelt, but they could [with all due respect] insist.

The magnitude of the moment was beginning to take its toll on Edward Smith. The Captain was having trouble making decisions. He did not issue a general order for evacuation. After his inspection tour with Andrews, he neglected to tell his officers the plight of their ship. An hour after the collision, Fourth Officer Boxhall was still unaware that she was sinking. Lightoller and Murdoch had to wait for permission to first fill the boats and then lower away. He also neglected to tell his officers that there were not enough lifeboats for all aboard. After the fact, Doctors would speculate that his actions indicated the Captain had suffered a nervous collapse.

As for the jewels in the Purser's safe, most left with their owners. With the ship starting to list, many of the ladies in First Class headed for McElroy's office to fetch their *"baubles."* McElroy had already sent crewmen to remind passengers to retrieve their valuables. When his safe was salvaged and opened in 1987, all that remained was a lone diamond necklace. One last time he had been equal to the task.

Soon the grand staircase was crowded with men and women heading [some reluctantly] upstairs to the Boat Deck. They moved on the red carpet, under the stained glass skylight, and past the clock depicting *"Honour and Glory crowning Time."* Mrs. Albert Caldwell moved along, as did the Astors and Jack Thayer. Master Thayer had good reason to step lively. Thomas Andrews had just told his father John, *"that he didn't give the ship much over an hour to live."*

Chapter 8
Twenty Boats

"The sea speaks a language
polite people never repeat.
It is colossal scavenger slang
and has no respect."

Carl Sandburg

By 12:30am, the First and Second class passengers were, for the most part, assembled on the Boat Deck, although many took refuge from the cold inside the public rooms. The Astors [John in a blue serge suit and Madeleine, nicely turned out in a light colored dress] were in the gym. As they sat astride two mechanical horses, Mr. Astor took a penknife and cut open a lifebelt, to show the contents to his bride.

On the Boat Deck, there was no panic, just a little concern and a lot of confusion, all of this accompanied by the ear-shattering sound of safety valves venting steam. As they stood there in the cold, they watched as the crew cleared away the lifeboats. There were sixteen wooden boats, eight to a side. They were arranged in clusters of four, two toward the bow and two toward the stern, and all hung from Welin Davits. The four Engelhart collapsibles were forward. "C" and "D" were inboard of the cutters, [Boats No. 1 and No. 2] while "A" and "B" were on the roof of the officer's quarters. Portside boats had even numbers, starboard side the odds, and were numbered in sequence bow to stern. Harland & Wolff, simultaneously with the *Titanic* and *Olympic,* built the fourteen main lifeboats and the two cutters. They had been designed for maximum seaworthiness.

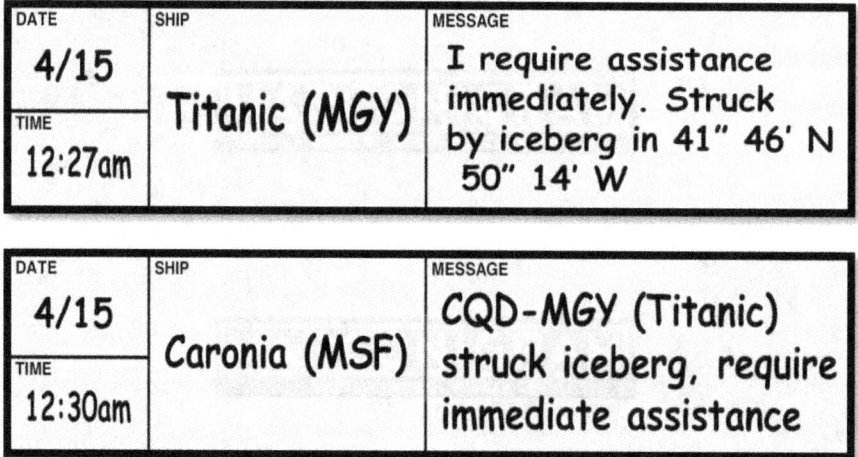

DATE	SHIP	MESSAGE
4/15	Titanic (MGY)	I require assistance immediately. Struck by iceberg in 41" 46' N 50" 14' W
TIME		
12:27am		

DATE	SHIP	MESSAGE
4/15	Caronia (MSF)	CQD-MGY (Titanic) struck iceberg, require immediate assistance
TIME		
12:30am		

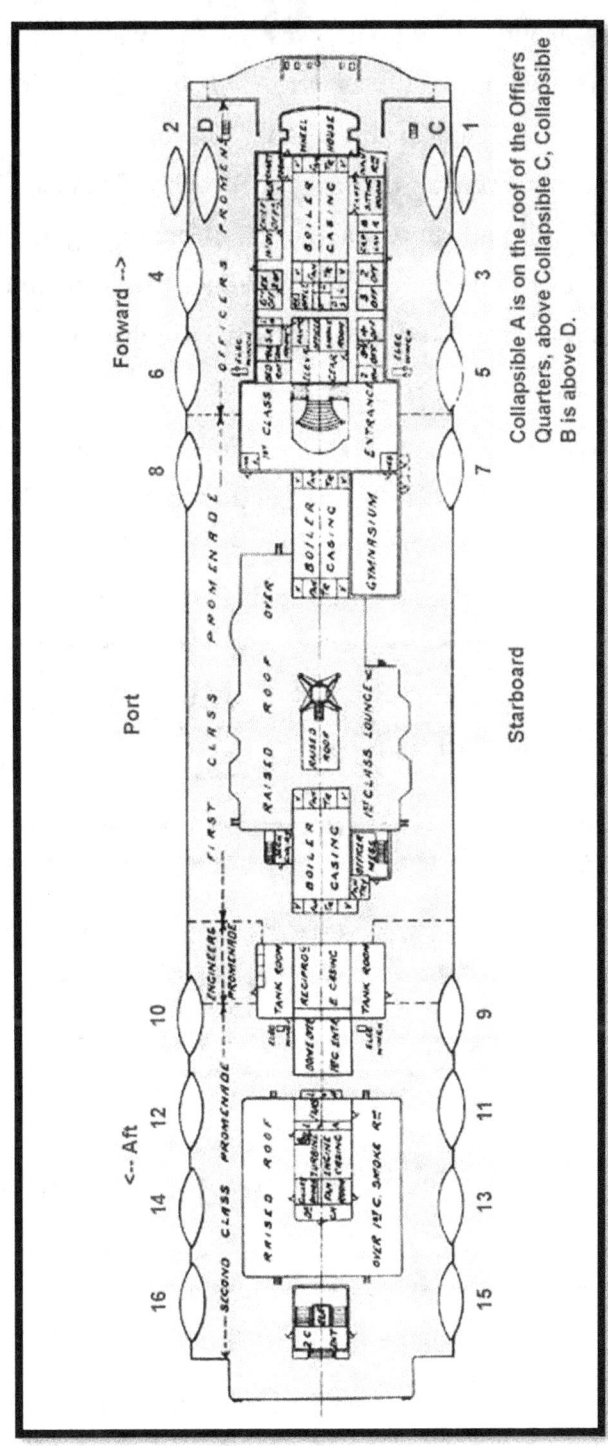

The Boat Deck

The fencing around the boats was taken away and the covers removed. The equipment was arranged and then with crewmen using spanners [wrenches] and cranks the boats were swung out. This was not, however, an exercise in British efficiency. There had been no boat drill and there was no list of boat assignments for the passengers. The majority of the crew had, in all likelihood, never been trained on nor worked a Welin Davit. Hugh Woolner would tell the U. S. Senate Inquiry, *"It struck me as being rather a slow job."*

As the night wore on, more ships came into range. Some had received the word while others most definitely had not. Witness this exchange between Phillips and the North German Lloyd's SS Frankfurt. It would not be their last conversation of the evening.

DATE	SHIP	MESSAGE
4/15	Frankfurt (DFT)	My position 39' 47" N, 52' 10" W.
TIME 12:34am		

DATE	SHIP	MESSAGE
4/15	Titanic (MGY)	Are you coming to our assistance?
TIME 12:34am		

DATE	SHIP	MESSAGE
4/15	Frankfurt (DFT)	What is the matter with you?
TIME 12:34am		

DATE	SHIP	MESSAGE
4/15	Titanic (MGY)	We have struck an iceberg and sinking. Please tell Captain to come.
TIME 12:34am		

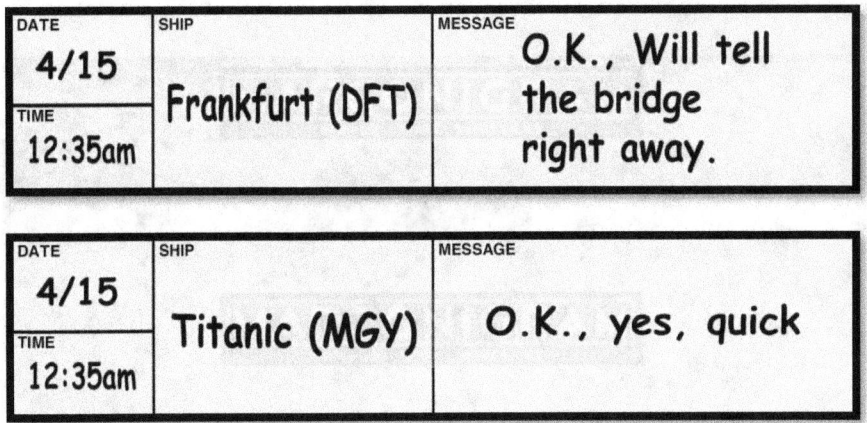

The boats would be lowered in sequence, from the middle forward then aft, with First Officer William Murdoch, Third Officer Herbert Pitman and Fifth Officer Harold Lowe working on the starboard side. Chief Officer Henry Wilde and Second Officer Charles Lightoller handled the port side with the assistance of Captain Edward Smith and Fourth Officer Joseph Boxhall. The time was 12:35 and Murdoch was standing by a now ready Lifeboat No. 7. As per his Captain's instructions, he called for women and children.

Convincing wary woman and their children to give up the solid deck, or a public room of the *Titanic,* for a sixty-five foot drop to the ocean surface in freezing weather in a little lifeboat could tax the talents of any officer, even one of Murdoch's ability. To many passengers their ship, even though now down by the bow, still seemed far safer than a rickety wooden boat. His duty, however, was clear and the First Officer got on with it. Dorothy Gibson, whose new movie, "The Lucky Hold Up," premiered the day after *Titanic* sailed from Southampton, came forward along with her mother Pauline. After searching in vain for her husband, Catherine Crosby climbed in with her daughter Harriette. New York heiress Margaret Hays got in, along with her dog *"Lady."* A handful of other First Class ladies entered the boat.

Dorothy Gibson, *"The Lucky Hold Up"*

At this point we need to address the lifeboat situation. First we'll take up *"Who?"* Captain Smith's direct order to Murdoch and Lightoller was, *"put the women and children in and lower away."* As you will see, the Second Officer followed his orders to the letter. On Murdoch's side things were different, especially regarding the first few boats. If the women would not get in, or get in alone, he allowed gentlemen into the boats. Honeymooners Dick and Helen Bishop [without her dog, *"Freu Freu"*] climbed in together. Mrs. Gibson's bridge partners, New York bankers Frederick K. Seward and William T. Sloper [remember that name] also came aboard. James McGough [he of the roomful of ice] and French aviator Pierre Maréchal were also permitted to board. Murdoch assigned lookouts George Hogg and Archie Jewel to the boat, and started to lower away.

And that brings us to "How?" Fifth Officer Lowe later testified at the U.S. Senate Inquiry that the ship's officers believed that the lifeboats were at risk of buckling and breaking apart if they were lowered from the davits full. They had intended that once the boats reached the water they would load passengers from doors in the ship's side or pick up swimmers in the water. First of all, no doors were opened into the side of the ship and only one boat picked up any survivors following the wreck [Lowe himself and No. 14]. Edward Wilding, representing Harland & Wolff, testified that the lifeboats had, in fact been tested at the shipyard with the equivalent of a full load of passengers and were lowered safely. The results of the tests had not been passed on to the officers of the *Titanic*. So Murdoch, believing he was acting with prudence, launched No. 7. A boat with a capacity of 65 entered the water with 28 passengers and crew. The time was 12:40am. But the saga of No. 7 was not over yet. In the bottom of each lifeboat was a hole used to drain water out while it's on the deck.

The drain hole had a plug that was to be inserted before a boat was launched. As the craft reached the surface, seawater began gushing from the opening. The lifeboat was lowered either without its plug or with the plug displaced. The problem was remedied by volunteer contributions from the lingerie of the women and the garments of the men. Regardless, those aboard had to sit for hours with their feet and ankles soaking in ice-cold water.

Murdoch and Lowe, now joined by Third Officer Pittman, immediately moved forward to No. 5. Also showing up to "help" was J. Bruce Ismay. Most of the passengers, unaware of the seriousness of their situation, were still reluctant to get into a lifeboat, in spite of Captain Smith shouting to one and all, *"Put on your lifebelts!"* Astor remarked, "We are safer on board the ship than in that little boat."

Mrs. Sallie Beckwith, her husband Richard and daughter Helen were in front of No. 5, part of a group that included a tennis player named Karl Behr [Mrs. Beckwith's *"most ardent admirer"*], Helen Newsom, and Edwin and Gertrude Kimball. Although the Third Officer was in charge of loading No. 5, Ismay was also urging wary passengers into the boat. Mrs. Kimball stepped forward and asked if they could all go together, and Ismay replied, *"Of course, madam, every one of you."* As a group, they entered Lifeboat No. 5.

New York Engineer Norman Chambers and his wife Bertha, Mrs. Eleanor Cassebeer, and Edward Calderhead, the buyer for the toy department at *"Gimbel's,"* joined them. Dr. Washington Dodge helped his wife Ruth and their four-year-old son into the boat, but made no attempt to board himself. As he watched, Dodge felt *"overwhelmed with doubts"* that he might be subjecting his family to greater danger in the lifeboat than if they had remained on *Titanic*.

Charles Henry Stengel saw his wife Annie into the boat and then walked away. He missed what followed. His wife was knocked unconscious and had two ribs broken when a 250 lb. passenger, Dr. Henry Frauenthal, jumped on top of her and into the lifeboat just before she was lowered.

Ismay sought to spur those lowering the boat to greater urgency by calling out repeatedly, *"Lower away!"* If his intent was to incur the ire of Lieutenant Lowe, he succeeded mightily, *"If you'll get the hell out of the way, I'll be able to do something! You want me to lower away quickly? You'll have me drown the lot of them!"* Ismay, having been properly told off by a junior officer, retreated up the deck, leaving the deck crew to wonder if, perhaps, the Fifth Officer didn't know to whom he was speaking. Murdoch then put Third Officer Pitman in charge and he ordered, *"Lower away!"*

As with the other boats, lowering No. 5 was a challenge. There was nothing smooth about it. First one side would jerk and drop a foot, and then the other would do the same. The equipment was too new, covered with paint and not properly oiled. The Welins came supplied with another improvement, power hoists for all the boats. None were ever used. In spite of it all, Pittman's Boat was away at 12:43am.

In the wireless room Bride and Phillips were trying to raise the *Olympic*. She was 500 miles away but her equipment was identical to her sisters - the finest on the water. Phillips has been sending *"CQD"* since 12:15am. Bride suggested, half jokingly, that he should try the new distress call, *"it might be your last chance to send it."* They all started laughing, even the Captain. So, at 12:45am on the morning of April 15, 1912, John Phillips, from the Marconi room of the *RMS Titanic,* sent what is believed to be history's first ever *"SOS."*

Jack Phillips, Senior Wireless Operator, *RMS Titanic*

Harold Bride, Junior Wireless Operator, *RMS Titanic*

DATE	SHIP	MESSAGE
4/15	Titanic (MGY)	SOS! SOS!
TIME 12:45am		41" 46' N 50" 14' W

In the steam bath that was Boiler Room No. 5, things were under control. The pumps had been rigged and were working; the fires had been drawn and assistant engineer Herbert Harvey has sent some of the stokers topside to their boat stations. Nothing to deal with until second assistant engineer Jonathan Shepherd stepped into a manhole and broke his leg. All seemed normal until the bulkhead between Boiler Rooms 6 and 5 gave way, and the North Atlantic came flooding in. The few hands remaining raced for the escape ladders. All but two made it out, Harvey and Shepherd. Nine of *Titanic's* twenty-four working boilers had been extinguished forever.

The light that Boxhall saw was clearly visible from the bridge. Neither the wireless nor the Morse Lamp were able to get a reply from the mystery ship. There was another way. QM George Rowe was on duty at the aft docking bridge and had not communicated with anyone for almost an hour when he telephoned the bridge. Rowe told the Fourth Officer he had seen a lifeboat [No. 7] in the water. Boxhall instructed the QM to find the pyrotechnics and bring them to the bridge. With permission from the Captain, Boxhall and Rowe sent up the first rocket at about 12:45am. *Titanic* had called "CQD," then "SOS," used her Morse Lamp and was now firing rockets.

> *"with a gasping sigh, one word escaped the lips of the crowd, rockets. Anybody knows what rockets at sea mean."*

The rocket burst a hundred feet above *Titanic's* deck. At the same time, mystery writer Jacques Futrelle arrived on the Boat Deck.

Futrelle, as well as some others, asked Second Officer Lightoller, *"Why are you getting the boats out?"* he had assured them, and himself, that the launchings were merely a precaution.

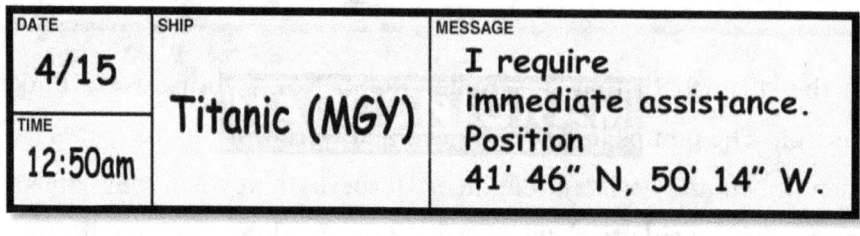

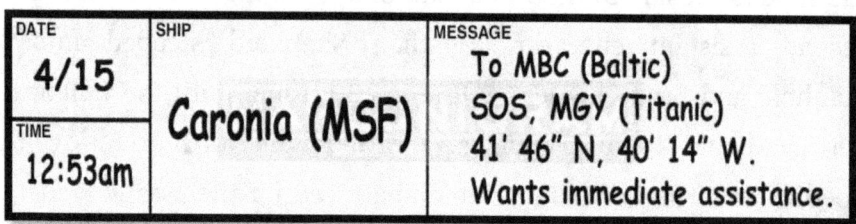

Next in line was boat No. 3. By then a pattern was developing. In the finest Edwardian tradition, male passengers helped their wives and children to board and then stood back. *"Women and children first"* meant something to these men, a call to duty, and they were prepared to go down with the ship. Charles Melville Hays and his wife Clara stood by on the Boat Deck. With them were their daughter Orian Davidson and her husband Thornton Davidson. The gentlemen saw their ladies into the boat and then made no attempt to board themselves. Margaret Brown later described the scene in an interview with *The New York Times*:

> *"The whole thing was so formal that it was difficult for anyone to realize it was a tragedy. Men and women stood in little groups and talked. Some laughed as the boats went over the side. All the time the band was playing. I can see the men up on deck tucking in the women and smiling. It was a strange night.*

It all seemed like a play like a dream that was being executed for entertainment. It did not seem real. Men would say 'After you' as they made some woman comfortable and stepped back."

Charlotte Drake Cardeza was returning from a Safari with her son and two servants in Promenade Suite B51/53/55. All four were escorted into No. 3. Once again Murdoch allowed men into the boat. Myra Harper boarded with her husband Henry Sleeper Harper, his dragoman, and their Pekinese dog "*Sun Yat Sen.*" Daisy Speeden climbed in with her son, and then moments later, her millionaire husband, Frederic Oakley Spedden. Albert and Vera Dick, who had been married on the same day the *Titanic* was launched, followed them. A handful of the ship's firemen were allowed to jump in. Murdoch put Able seaman George Moore in command of the boat and then lowered away. No. 3 suffered the same problems with lowering that the others had encountered. The lifeboat descending in fits and starts as the lowering ropes repeatedly stuck in the pulleys, but eventually reached the water safely. There were fourteen women in the boat and eighteen men. Of the men, eleven were members of the crew. The time was now 12:55am.

Aboard the *Californian,* it had been a half-hour since Second Officer Herbert Stone took over the watch from Third Officer C. V. Groves. On his watch Groves had tried signaling the ship with the Morse Lamp without success. Around 12:45am, Stone saw a white flash appear from the direction of the nearby ship. First he thought it was a shooting star until he saw another one. He saw five rockets before being joined on the bridge by apprentice crewman James Gibson. He called down the speaking tube to Captain Lord at 1:15am, but it is unclear how many rockets he told him about.

Lord asked, *"if they had been a company signal?"* Stone said he didn't know. The Captain told his Second Officer to call him if anything about the ship changed and to keep signaling with the Morse Lamp. Again, he made no request to wake up Cyril Evans and try to contact the steamer by wireless. On board the *Titanic,* Phillips and Bride had finally made contact with her sister.

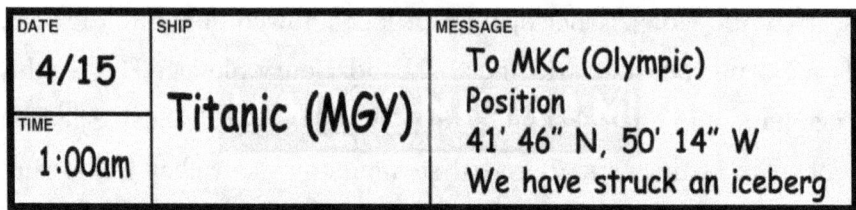

A few feet below the wireless cabin on the Boat Deck, Second Officer Lightoller, with the help of Captain Smith and Chief Officer Wilde, had No. 8 ready for loading. By now, Andrews was going down the line from boat to boat, urging women to get in. A group was gathered around the boat, preparing to board. Ida Straus was asked to join them but refused, *"I will not be separated from my husband, as we have lived, so will we die...together."* The 67-year-old Isidor likewise refused an offer to board on account of his age, saying, *"I do not wish any distinction in my favor which is not granted to others."* A moment later they went below, presumably back to their cabin. Neither Isidor nor Ida would ever be seen again.

Now, with one foot on the deck and the other in the boat, Lightoller asked for women and children only. Howard Case, London manager of Vacuum Oil, and steel heir Augustus Roebling helped Mrs. William T. Graham, her daughter Margaret and her governess into the boat. Emma Eliza Bucknell, whose husband founded Bucknell College, boarded with her maid, followed by The Countess of Rothes, cousin Gladys Cheery, and maid Roberta Maioni.

And while Mr. & Mrs. Straus did not enter the boat, their maid, Ellen Bird did. Along with encouragement from her employers, Ida gave Ellen her fur coat saying, *"I will not be needing it."*

Major Butt brought to the boat Marie Young, who had been a governess to the children of his friend President Theodore Roosevelt. She later recalled that he *"wrapped blankets about me and tucked me in as carefully as if we were going on a motor ride."* He wished her farewell and good luck, and asked her, *"don't forget to remember me to the folks back home."* There were no other women around, and Lightoller steadfastly refused to allow men to board. He put Able-seaman Thomas Jones in charge of the boat and lowered away. The Second Officer's pulleys and ropes were no better than Murdoch's. The boat lurched its way down, waisting precious time. As No. 8 dropped to the sea, Mrs. Graham watched Case leaning against the rail, light a cigarette and wave good-bye. Edith Graham was one of 25 people in the boat. The time was 1:00am.

On the bridge, the Captain knew the mystery ship could plainly see Boxhall's rockets. She looked so close. [So close in fact that Lifeboat No. 8 was ordered to row over to the ship, off load its passengers and come back for more.] As Rowe fired the rocket gun, Smith asked him if he could send and receive Morse. The QM replied that he could, *"a little."* *"Good, now call that ship up and when she replies, tell her we are the Titanic sinking; please have all your boats ready!"*

Still 58 miles away, *RMS Carpathia* sprang to life. The wireless operator and the First Officer, not bothering to knock, burst into Rostron's cabin, and before he could reprimand them, Cottam blurted out the news about the *Titanic*. *"Are you sure?"* *"Yes,"* came the reply. The Captain immediately ordered the ship to come about.

Fortunino Matania portrait, Loading lifeboats, *RMS Titanic*

Still in his pajamas, the Captain fixed her position on the chart, set a course for the stricken liner, and then sent for all his department heads. The Chief Engineer arrived first. Rostron ordered him to call another watch of stokers and make all possible speed. The First Officer was told to knock off all routine work and prepare for orders. Like a Vickers machine gun he started spitting them out:

English doctor, with assistants, to remain in First Class dining room.

Italian doctor, with assistants, to remain in Second Class dining room.

Hungarian doctor, with assistants, to remain in third class dining room.

Each doctor to have supplies of restoratives, stimulants, and everything to hand for immediate needs of wounded or sick.

Purser, with assistant purser and chief steward, to receive the passengers, etc., at different gangways, controlling our own stewards in assisting Titanic passengers to the dining rooms, etc.; also to get Christian and surnames of all survivors as soon as possible to send by wireless.

Inspector, steerage stewards, and master at arms to control our own steerage passengers and keep them out of the third class dining hall, and also to keep them out of the way and off the deck to prevent confusion.

Chief steward: That all hands would be called and to have coffee, etc., ready to serve out to all our crew.

Have coffee, tea, soup, etc., in each saloon, blankets in saloons, at the gangways, and some for the boats.

To see all rescued cared for and there wants attended to.

All officials' cabins to be given up. Smoke rooms, library, etc., would be utilized to accommodate the survivors.

All spare berths in steerage to be utilized for Titanic's passengers, and get all our own steerage passengers grouped together.

Stewards to be placed in each alleyway to reassure our own passengers, should they inquire about noise in getting our boats out, etc., or the working of engines.

Remind the crew of the necessity for order, discipline and quietness and to avoid all confusion.

Chief and first officers: All the hands to be called; get coffee, etc. Prepare and swing out all boats.

All gangway doors to be opened.

Electric sprays in each gangway and over side.

A block with line rove hooked in each gangway.

A chair sling at each gangway, for getting up sick.

Boatswains' chairs, Pilot ladders and canvas ash bags to be at each gangway, the canvas ash bags for children.

Cargo falls with both ends clear; bowlines in the ends, and bights secured along ship's sides, for boat ropes or to help the people up.

Heaving lines distributed along the ship's side, and gaskets handy near gangways for lashing people in chairs, etc.

Forward derricks, topped and rigged, and steam on winches; also told off officers for different stations and for certain eventualities.

Company's rockets to be fired at 2:45am and every quarter of an hour after to reassure Titanic.

The *"Electric Spark"* was firing on all cylinders.

First Officer Murdoch, having lowered the No. 3 Boat safely, now moved forward to No. 1. This boat was a cutter, an emergency boat. Since it had to be ready to go at a moments notice, it was already "swung out." It was smaller and more agile than a standard boat and had room for 40, as opposed to the bigger boats 65. As the men prepared to load her, Sir Cosmo Duff-Gordon asked if his party could board. Murdoch replied, *"Oh certainly do; I'd be very pleased."* We have no way of knowing if he was being sincere or sarcastic. At any rate, Sir Cosmo, his wife Lady Duff Gordon and her secretary, Laura Francatelli, were the first three people in the cutter. Abraham Lincoln Salomon was also allowed to enter, as was Charles Henry Stengel. Stengal had just seen his wife off in the No. 5 Boat. As he tried to enter the boat, he was unable to clear the bulwark railing, and thus had to "roll" into the boat, causing the First Officer to remark, *"This is the funniest thing I've seen all night."*

Six stokers had made their way from the boiler rooms up to the Boat Deck and were told to get in. Murdoch gave lookout George Symons command and ordered him to, *"Stand off the ship's side and return when we call you."* He then gave the signal to lower away. As she was being lowered with her dozen passengers, Greaser Walter Hurst remarked, *"If they are sending the boats away they might as well put some people in them."* Lifeboat No. 1, with room for 40, did indeed float into the night carrying 12 people. Controversy would follow in her wake. It was 1:05am.

While Bride and Phillips kept signaling with the Wireless, the stokers and trimmers kept up the fires in the remaining boiler rooms, the Engineers monitored the power supply and kept the steam pressure up, Boxhall and Rowe continued firing their rockets, and the deck hands labored mightily to clear away and load the boats.

In spite of an air temperature of 28° Fahrenheit, the men were sweating profusely, none more than the Second Officer. Lightoller removed his coat and continued his labors, now clad in just a sweater and his trousers. Seeing this, Doctor O'Loughlin [forever the wag] called out to him, *"Hello 'Lights,' are you warm?"*

Colonel Archibald Gracie, First Class Passenger

Major Arthur Peuchen, First Class passenger

Of the twenty boats that were launched or floated off the *Titanic*, probably the most famous was Lifeboat No. 6. It was shortly after 1:00am that Lightoller had her ready to board, and by now his *"women and children only"* policy was up and running. Of the 23 people who would leave the *Titanic* in No. 6, 19 would be women.

Colonel Gracie, the self-anointed "protector" of the unescorted women in First Class, was at his wits end. None of the women he swore to help could be found, especially the *"Jewel in the Crown,"* Helen Candee. Fellow *our coterie* members, architect Edward Kent and sculptor Hugh Woolner, were on the case. Woolner reached her first, and was escorting Helen up the stairs when Kent arrived. Fearing for Kent's life, she handed him a treasured cameo of her mother, making him promise to return it to her after they were rescued. Her friends helped her into No. 6., with Woolner assuring her that they would help her on board when the *Titanic* steadied herself. Moments later Gracie arrived, still looking for Helen. Woolner told his friend, rather smugly that he had looked after her and that she was safely away, albeit with a broken ankle suffered while boarding. [Of her suitors, many believed it was Woolner that Helen fancied. Four months later he would marry - someone else.]

With Quartermaster Rowe's distress rockets exploding overhead, it was getting easier to convince people into the boats. Still some still did not go willingly. Mary E. Smith had refused to get into Lifeboat No. 8, but was forced by her husband into No. 6 with the words, *"Keep your hands in your pockets, it's very cold weather!"* Quigg Baxter rushed up with his mother, sister and his lover. Bertha Mayné had slipped a woolen overcoat over her nightdress but balked when Quigg insisted that she get in the boat without him. She wanted to go back to her cabin, but Maggie Brown talked her out of it.

Mrs. Elizabeth Rothschild, wife of the New York clothing manufacturer, climbed in and somehow managed to slip her Pomeranian past Lightoller. Mrs. Martha Evelyn Stone who, in the passenger manifest listed her address as *The Plaza Hotel* boarded with her maid. Even the intrepid Mrs. Brown did not board voluntarily. It was left to a crewman to pick "Maggie" up and drop her bodily into the boat as it was being lowered.

Lightoller gave the tiller to Quartermaster Robert Hichens and added lookout Frederick Fleet. As the boat neared the water, someone shouted up to the deck. *"We have only one man to row!"* Stories vary on who did the shouting, some say it was Hichens; others claim it was Mrs. Brown. Lightoller appealed to the crowd still on deck for anyone who had sailing experience. Major Arthur Godfrey Peuchen of the Royal Canadian Yacht Club volunteered, telling the Second Officer that he was a yachtsman. Captain Smith, standing nearby, suggested Peuchen should go down one flight and break a window on the Promenade Deck to get into the boat. Lightoller replied, *"that if he was as good a sailor as he claimed to be, he could slip down the ropes to get into the boat."* So Peuchen grabbed a fall, swung himself off the ship, and hand under hand slithered down 30 feet of rope into the boat. The Major had the distinction of being the only man Lightoller would allow in a lifeboat. No. 6, now with a second gentleman to man the oars, got away safely at 1:10am. In the wireless room, Phillips and Bride added another word to their calls, *"sinking."*

DATE	SHIP	MESSAGE
4/15	Titanic (MGY)	Titanic to MKC (Olympic) We are in collision with berg. Sinking Head down. 41.46 N. 50.14 W. Come soon as possible.
TIME		
1:10am		

DATE	SHIP	MESSAGE
4/15	Titanic (MGY)	Titanic to MKC (Olympic) Captain says, "Get your boats ready. What is your position?"
TIME		
1:10am		

The *Olympic's* position put her almost 500 miles away from her sister [a full days steaming]. With the water line coming closer and the slant of the deck increasing, very little persuasion was now needed to fill the boats. With Lightoller still engaged with the No. 6 Boat, Lifeboat No. 16 became the responsibility of Sixth Officer Moody. His boat was in the aft cluster on the starboard side. This was Second Class promenade space. The area was filling with Second and Third Class passengers, who made their way up to the Boat Deck through First Class. Stewards and boatswains had been posted in passageways, trying to keep the steerage travelers in Third. No doubt that on C or B-Deck there was a steward or two who had been knocked into unconscious by husbands who were bound and determined to get their wives and children up to the boats.

Moody now asked for women. Carla Jensen boarded, leaving behind her uncle, brother and fiancé. Agnes McCoy, who was relocating to Brooklyn, climbed in with her sister Alice. Surprisingly, Moody allowed their 24-year-old brother Bernard to join them. Her husband Thomas escorted Mary Davidson to the boat. Two days from now her parents, Mr. And Mrs. Fink of Cleveland, Ohio, would receive a letter from their daughter telling them that she and Thomas had booked passage on the *Titanic*. Adolf Dyker helped his wife Anna into the craft. He kissed her and stepped back to let other women enter the boat. In her handbag, among other things, were two golden watches, two diamond rings and a sapphire necklace.

James Paul Moody, Sixth Officer, RMS Titanic

Edwina Troutt, Second Class Passenger, *RMS Titanic* **(1955)**

This much jewelry was typical for a First Class passenger, but the Dykers were traveling in steerage. The Murphy girls, 25-year old Maggie and 18-year old Kate, had boarded together. They had been stopped below by a seaman but were allowed to pass when passenger James Farrell yelled, *"Great God, man! Open the gate and let the girls through!"* Farrell, a farm laborer, would perish, but Kate would never forget him. They were followed by Evelyn Marsden; a Stewardess whose previous ship had been the *Olympic*. She signed on with the *Titanic* four days before the liner left Southampton.

Violet Jessop could be excused if she saw herself as a jinx. First she endured the collision on the *Olympic* and now this. Moody recognized her and ordered that she get into the boat to show the other women that it was safe. As the boat was being lowered, the Sixth Officer called, *"Here, Miss Jessop. Look after this baby."* And a bundle was dropped in her lap." The stewardess was not the only woman aboard to be "blessed." One of the last passengers to load was Winnie Trout. As she waited for the boat to be lowered, a Lebanese passenger, Charles Thomas, came past with his nephew. He begged for the infant to be saved and Winnie took the child into the lifeboat. As the boat started down, she clutched a toothbrush, a prayer book and the five-month-old child. Moody put Master-at-Arms Henry Joseph Bailey in charge, rounded up five other members of the crew, and lowered away. Approximately 50 people were aboard, most of them women from Third Class. Lifeboat No. 16 touched the surface at 1:20am.

As he finished loading No. 6, Wilde came to Lightoller to ask where the firearms were kept. They had been his responsibility when he was the First Officer in Belfast. *"Lights"* led Captain Smith, Wilde and the other officers [including McElroy] to the locker in the First Officer's cabin. As they were about to leave, Wilde shoved a Webley .455 caliber revolver and ammunition into his hand. *"You might need this."* Lightoller was doubtful, but events would prove the Chief Officer right.

Wilde, Lightoller, Moody and Lowe left to load Lifeboat No. 14. The *Titanic* was now well down by the head and some of the passengers, fearing the worst, were beginning to panic. As they pressed forward, Fifth Officer Lowe took out his sidearm and fired three shots in the air, in an effort to get them back. Mrs. Biela Moor and her seven-year-old son, transfers from the *Adriatic,* having been jostled up the stairway by the surging crowd, made it safely into the boat. To be followed by the Collyers, mother [enduring heartburn] and child.

Mrs. Esther Hart had not slept since she boarded the ship in Southampton, tormented by the thought that some kind of catastrophe would hit the new liner. To call a ship *"unsinkable"* was, in her mind, *"flying in the face of God."* Now with her worst fear realized, she climbed into the lifeboat with her seven-year-old daughter Eva. Her husband Benjamin had wrapped Eva in a blanket and carried her to the Boat Deck. As he placed his daughter in the craft he told her to, *"hold mummy's hand and be a good girl."*

Mrs. Elizabeth Davies was there with her three children. Her eldest, Joseph, helped place his mother, brother and sister in the boat and then asked if he could join them. Lowe's refusal came bundled with a threat. He would shoot the boy if he attempted to get in.

Harold Godfrey Lowe, Fifth Officer, *RMS Titanic*

Benjamin, Eva, and Esther Hart, *RMS Titanic*

Mrs. Davies pleaded with the Fifth Officer that her son be allowed to come aboard, that there was plenty of room in the boat. It was, however, all for naught. Lowe and Moody felt this boat needed an officer. The "rules" said that the junior officer should go, but Moody insisted that Lowe take command. The Fifth Officer took the tiller and then ordered *"Lower Away!"* The decision would cost the Sixth Officer his life, but would pay huge dividends in the hours to come.

As the boat was lowered, a young man climbed over the rails and tried to hide under the seats. Lowe ordered him to leave at gunpoint, first threatening to *"blow your brains out,"* and then appealing to him to *"be a man, we've got women and children to save."* By now the women in the boat were crying. Marjory Collyer, Charlotte's eight-year-old daughter started tugging on Lowe's arm and pleading, *"Oh Mr. man, don't shoot, please don't shoot the poor man!"* The gentleman was returned to the Boat Deck, there to await his fate.

Now a group of men rushed the boat. Seaman Joseph Scarrott had to beat them back with the tiller. He would later say that since they didn't understand him, *"they must have been foreigners."* [Foreigners, especially Italians, would be blamed for everything.] That was enough for Lowe. He drew his Webley pistol. Again shots rang out. *"If anyone else tries that, this is what he will get,"* as he fired the revolver twice. That stopped the stampede. No. 14 [like the others] with room for 65, continued down to the water with 40 on board. The clock above the grand staircase read 1:25am.

DATE	SHIP	MESSAGE
4/15	Caronia (MSF)	To MGY (Titanic) "Baltic coming to your assistance."
TIME		
1:25am		

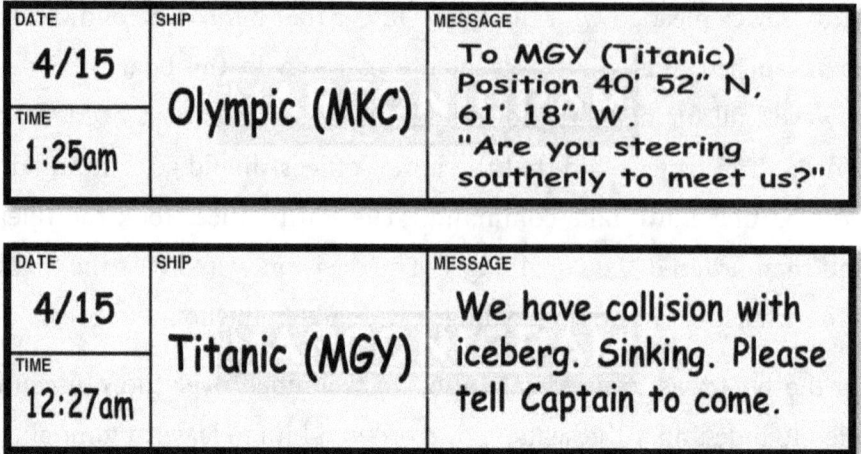

By now the water was starting to seep into Boiler Room No. 4. The fires had to be drawn to prevent what *"the black gang"* feared most, a boiler explosion. The loss of this room meant there was now less power available to run the pumps, keep the lights on and power the wireless. To save power all the ventilation fans in the room were turned off. The water level continued to rise and when it reached the engineers knees, No. 4 was abandoned. The watertight doors, which had been opened for access to the room, were now closed again. It was a matter of time before the water reached Boiler Room No. 3.

Time was running out and Lightoller knew it. A look down the emergency stairwell at the rising tide was proof enough for him. He and Wilde moved on to Lifeboat No. 12. There were fewer crewman left on deck now and only two were available for this boat. Able-Seaman John Poigndestre was put in charge and was assisted by Able-Seaman Frederick Clench. Forty people were loaded, mostly women from Second and Third Class. As the boat was being lowered past B-Deck, an unidentified male passenger jumped in. The boat reached the water without incident but then a problem arose. The crewmen were unable to detach the falls from the boat.

After several attempts failed to release the vessel, Poigndestre produced a knife from his pocket and cut through the ropes. Lifeboat No. 12 drifted off with 42 people aboard, at 1:30am. In the wireless room, the ever vigilant Phillips and Bride kept sending reports.

DATE 4/15 TIME 1:30am	SHIP Titanic (MGY)	MESSAGE To MKC (Olympic) Women and children in boats, can not last much longer.

Across the Boat Deck, Murdoch, with help from Chief Purser McElroy was preparing Lifeboat No. 9. Among the first to board was Léontine Aubart, mistress of Benjamin Guggenheim, and her maid Emma Sägasser. Guggenheim saw her safely in the boat and then returned to his stateroom, accompanied by his valet, Victor Giglio.

Even at this late hour, some were reluctant to get into a boat. One elderly woman refused to board, making a great fuss, and retreated below decks. Lily Futrelle, the wife of novelist Jacques Futrelle, was likewise initially reluctant to board; but after her husband told her, *"For God's sake, go! It's your last chance! Go!"* An officer [probably Murdoch] forced her into the boat. Most of the passengers were women, with two or three men entering when no more ladies came forward. Two "gentlemen" from First Class got into the boat. Charles Romaine and George Brayton [and on occasion Brereton] had something in common. They were both professional gamblers.

DATE 4/15 TIME 1:35pm	SHIP Olympic (MKC)	MESSAGE What weather do you have?

DATE	SHIP	MESSAGE
4/15	Titanic (MGY)	To MKC (Olympic)
TIME		Clear and calm
1:35pm		

Kate Buss and her friend Marion Wright were standing with their shipboard acquaintances Douglas Norman and Dr. Alfred Pain, watching the boats being lowered, when a call came for *"Any more ladies?"* The two men brought Buss and Wright to Boat No. 9, and the ladies beckoned Norman and Pain to join them. However, the men were barred from entering by crewmen on the deck. Horrified, Buss demanded to know why they had not been allowed aboard. Boatswain's Mate Albert Haines told her why:

"The officer gave the order to lower away, and if I didn't do so he might shoot me, and simply put someone else in charge, and your friends would still not be allowed to come."

Benjamin Guggenheim was back on deck with his valet. His career had been nothing but a squandered fortune, his marriage to Florette nothing but a sham. He had led a privileged Edwardian life without honor, but, was determined the last chapter would be different. The millionaire and his body man, without lifebelts, stood resplendent in their evening clothes. Guggenheim confided to a steward:

"We've dressed in our best, and are prepared to go down like gentlemen. There is grave doubt that the men will get off. I am willing to remain and play the man's game if there are not enough boats for more than the women and children. I won't die here like a beast. Tell my wife I played the game out straight and to the end. No women shall be left aboard this ship because Ben Guggenheim was a coward."

While Guggenheim played out his version of *"noblesse oblige,"* the First Officer put Haines in charge, with Able-Seaman George McGough [no relation to James] at the tiller. Moments later, Boat No. 9 was on the water with 40 of her 65 seats occupied. It was now 1:35am. With the water level ever rising, Captain Smith returned to the wireless room to deliver still more bad news. Phillips immediately put it out over the air.

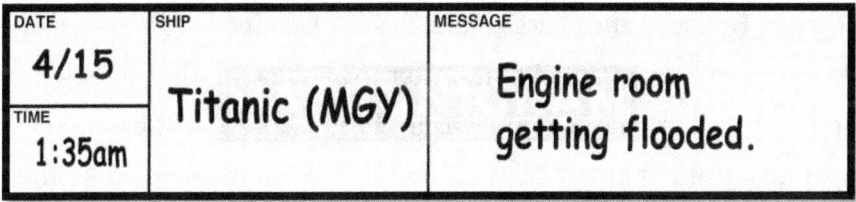

First Officer Murdoch then took over the filling of Lifeboat No. 11 with Able-seaman Sidney Humphreys in charge. By now the lifeboats were being filled much closer to their capacity.

Little tragedies were taking place all over the deck, many of them in and around the No. 11 Boat. Eighteen-year-old Leah Aks and her ten-month-old son Philip were headed for America. After the ship struck the iceberg, Leah and "Filly" followed other Third Class passengers to the bottom of the Third-Class staircase at the rear of the ship. At 12:30pm, the crew permitted women and children in this group to make their way to the Boat Deck. When crewmembers saw that Leah and "Filly" couldn't get through the crowd and up the stairs, they carried the two to the Boat Deck. Madeline Astor, herself an eighteen-year-old bride, covered the baby's head with her silk scarf. Then a crazed man, who had been refused entry by the crew when he attempted to get into a lifeboat, ran up to Leah and said, *"I'll show you women and children first!"* The man grabbed "Filly" and appeared to throw the infant overboard.

The distraught mother ran up and down the deck looking for her son. What she didn't know was that "Filly" had fallen into No. 11, right into another woman's arms. The woman is presumed to have been Italian immigrant Argene del Carlo. Her husband was not permitted to follow the pregnant Argene into the lifeboat.

"Filly" would be the youngest person in No. 11, by one month. Eleven-month-old Trevor Allison was returning to Quebec with his parents, Hudson and Bessie, and sister Loraine. A nurse, Alice Cleaver, also accompanied him. After the collision, Alice bundled up the infant in her charge and went up to the Boat Deck, having been separated from the rest of the family. Bedroom Steward William Faulkner held baby Trevor while Alice got in to the boat. Bess Allison was put in another boat [probably No. 6] with Loraine but refused to leave the ship without her baby. She dragged Loraine out of the boat and started searching for Alice and Trevor. When told her husband had the baby and were in a boat being lowered on the portside of the deck, she ran to the other side, only to find that Mr. Allison was not there. No. 11 would leave without Mrs. Allison or Loraine.

René Jacques Lévy, the gentleman who just yesterday made the boast *"he'd rather go down with the ship than get in a lifeboat,"* found his friend Marie Jerwan on B-Deck. He, together with cabin-mate Jean-Noël Malachard, told her, *"We'll take care of you."* They went to the Boat Deck and placed Marie Jerwan into the No. 11 Boat. The men shouted: *"Good-bye,"* as the boat was winched down and waved their hands. That was the last time Lévy or Malachard were seen.

Ruth Becker was traveling from India to the U. S. with her mother Nellie and two siblings. A steward placed her brother Richard and sister Marion into No. 11 and then said *"Well, that's all for this boat."*

At which point her mother pleaded to be allowed in as well. *"Please let me in this boat! Those are my children!"* She was finally allowed to board but Ruth was left on the Boat Deck as her mother screamed, *"Ruth! Get in another boat!"*

It wasn't, however, all tragic. Steward James Witter had not intended to board but was knocked into No. 11 by a hysterical woman whom he was helping. Edith Louise Rosenbaum brought along her lucky toy pig, which played the *"Maxixe."* She had wrapped the pig in a blanket to protect it but was too frightened to enter the lifeboat. Thinking it was a baby, a steward took it and tossed it to one of the women already aboard. Rosenbaum could not bear the thought of losing her precious pig and boarded the lifeboat to retrieve it.

Murdoch ordered the boat lowered, but No. 11 was not out of the woods, not just yet. As she reached the surface, the little boat was nearly swamped by high-pressure water being pumped out of *Titanic*. The now partially flooded boat moved away with some of its passengers having to stand knee deep in the North Atlantic. She had been filled beyond capacity with 70 people in a boat designed for 65. The chronometer in the ship's wheelhouse showed the time: 1:35am.

On the portside of the bridge, Boxhall and QM Rowe continued to fire rockets, hoping to get a response from the "mystery ship," but it was all for naught. A frustrated Fourth Officer wished that he had another option, *"If I had a cannon, I'd have put a shell in her side."*

DATE	SHIP	MESSAGE
4/15	Frankfurt (DFT)	To (MGY) Titanic Are there any boats around you already?
TIME		
1:35am		

Ruth Becker, Second Class passenger, *RMS Titanic*

Murdoch and Moody now moved aft to Lifeboat No. 13. The fear of filling a boat while still on the davits was gone. Thirteen and subsequent boats would be close to, at, or above capacity. The boat would be loaded first from the Boat Deck, and then from B-Deck, with predominantly Second and Third Class women and children. And yet, at the eleventh hour, there were still some who were reluctant to get in a boat. One woman was delirious, crying, *"Don't put me in that boat! I don't want to go in that boat! I've never been in an open boat in my life!"* Steward F. Dent Ray's ungracious response: *"You have got to go and you may as well keep quiet."* Anna DeMessemaeker became hysterical and would not leave her husband's side. He forcibly picked her up and handed her to an officer in No. 13. She sat in the boat, convinced that she would never see him again. Moody placed Ruth Becker in the boat after she was prevented from entering No. 11, with her mother, brother and sister.

Loading continued when the boat was lowered to B-Deck. Sylvia Mae Caldwell got into the boat. Her 10-month-old son Alden, who was wrapped in a blanket, was tossed to Steward Frederick Ray in the stern. Sylvia's husband Albert stepped into the bow. Murdoch, after having dragged some out of the boat, now allowed men to board. Second Class schoolteacher Lawrence Beesley climbed in along with Daniel Buckley. [His head covered with a shawl.] Dr. Washington Dodge was also allowed to board, having earlier seen his wife and child get away safely in the No. 5 Boat. He owed his presence aboard the boat to the apparent guilty feelings of Steward Ray, who had urged Dodge and his wife to sail on *Titanic* in the first place. Just before No. 13 was lowered, Ray bundled Dodge aboard. More stokers and trimmers from the boiler rooms were ordered in. Only Ruth Becker had the foresight to bring some blankets.

Immediately, she passed them out to the shirt-sleeved boiler room "refugees," already shivering in the cold. The First Officer put Leading Fireman Frederick Barrett in charge. She had a full load of 65 aboard when the order came to *"Lower Away!"*

DATE	SHIP	MESSAGE
4/15	Baltic (MBC)	To (MGY) Titanic
TIME		We are rushing
1:37am		to you.

While it was being lowered, the lifeboat was nearly caught by an enormous stream of water, three or four feet in diameter. This was coming from the condenser exhaust that was being produced by the pumps, trying to expel the water that was flooding *Titanic*. The occupants had to push the boat clear using their oars and spars but finally reached the water. They were safe, for the moment.

Murdoch and Moody managed to fill the No. 15 boat simultaneously with Lifeboat No. 13. The First Officer placed Fireman Frank Dyamond in charge of the boat. The officers, now determined to fill each craft to the brim, started loading.

Miss Bertha E. Mulvihill, a 24-year-old colleen, was coming home from a wedding in Ireland, on her way to Rhode Island with bridal plans of her own. With her coat on over nightdress, she some how made it from Third Class all the way to the boats.

Elin Hakkarainen had no time to dress properly, but she grabbed her handbag and lifebelt and hurried to the corridor. All the stairway gates appeared to be locked, but, at last, she noticed a steward coming to collect a group of steerage passengers. He guided them through the maze of passageways and up to the Boat Deck.

Elin searched for her husband, but in the crush of people could not find him. An officer told her that there was space for her in No. 15.

Eastbound from New York, *RMS Olympic* was twenty hours away, making 21.5 knots. With the coal strike aftermath still lingering, and no one sure what awaited her in Southampton, the liner was, like the *Titanic*, "managing" her fuel with several boilers shut down. But now, finally came the realization that her sister was in trouble, and to Captain Haddock, saving coal had been dropped off his priority list.

DATE	SHIP	MESSAGE
4/15	Olympic (MKC)	To MGY (Titanic) Am lighting up all possible boilers as fast as (we) can.
TIME		
1:40am		

Eighteen-year-old Anna Turja shared a Third Class cabin with a friend, her children and a neighbor. There was a knock on the door, a voice called, *"Get up or soon you will be at the bottom of the ocean!"* As they made they way towards the deck, a seaman tried to bar their way but Anna and her party refused to obey. The crewman didn't stop to argue with them but the doors were closed and chained behind their group to prevent others from coming up. Finding another route to the Boat Deck, she heard the music played by Wallace Hartley's orchestra and then was placed in the boat.

Oscar Hedman worked as a settler recruiter. His job was to find migrant workers and bring them to the United States. On this trip he was to bring seventeen [most of whom could not speak English] to work out west. On reaching the No. 15 boat, Hedman, having lost track of his immigrant charges, threw himself into the lifeboat at the last second. He would say later that at that moment he had thought to himself, *"If they are going to shoot me, I'll just die faster."*

Lifeboats No. 13 & No. 15, *RMS Titanic*

In all, the Officers managed to squeeze 68 into a boat with a capacity of 65. She would enter the water the second most crowded of all *Titanic's* lifeboats. That is, if she entered the water at all. Halfway down the lowering was abruptly stopped. People were screaming from the surface. As a result of the wash from the pumps, Lifeboat No. 13 had drifted under No. 15. To make matters worse, the falls aboard No. 13 jammed and had to be cut free to allow her to get away safely. The boat had come within a few feet of being swamped by the weight of the craft above her. Once she had moved away, lowering resumed, and No. 15 was free to reach the surface. Loaded beyond capacity, the gunwales [the top edge of a side of a boat] were so far down in the water, one female passenger later said that when she leaned against the side her hair trailed in the water.

At the other end of the ship, on the portside wing of the bridge, Boxhall and Quartermaster Rowe fired their 8th and final rocket. They were still trying to raise the ship they could both plainly see. Whoever was out there was still not responding. The Fourth Officer left to assist with loading the boats; Rowe remained and resumed trying to contact her with the Morse Lamp.

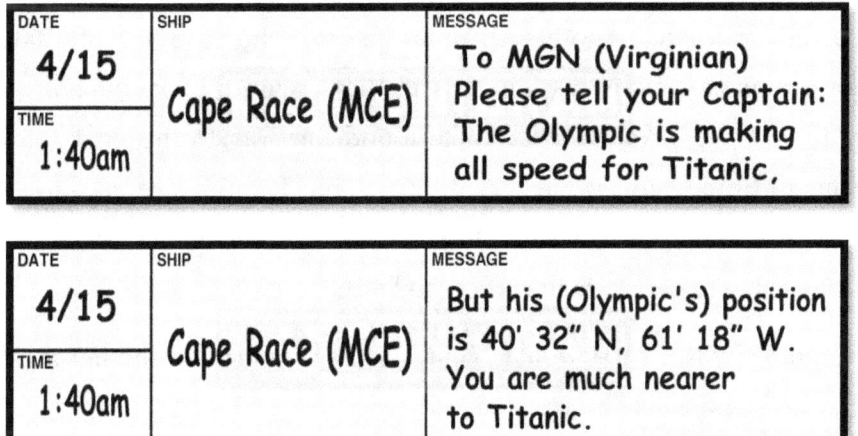

DATE	SHIP	MESSAGE
4/15	Cape Race (MCE)	To MGN (Virginian) Please tell your Captain: The Olympic is making all speed for Titanic,
TIME 1:40am		

DATE	SHIP	MESSAGE
4/15	Cape Race (MCE)	But his (Olympic's) position is 40' 32" N, 61' 18" W. You are much nearer to Titanic.
TIME 1:40am		

DATE	SHIP	MESSAGE
4/15	Cape Race (MCE)	The Titanic is already putting women off in the boats, and he says weather there is calm and clear.
TIME		
1:40am		

On the port side, no less than the Captain, the Chief Officer, and the Second Officer helped with Lifeboat No. 2. There was a higher feeling of urgency now. Not only was *Titanic's* bow almost to the waterline, but the cutter [No. 2] had to be lowered quickly so that an Englehardt collapsible [which shared the same davits] could be attached. Lightoller walked to the boat to prepare it for loading and found it already filled with a large group of male passengers and crewmen. He drew his gun, telling them, "*Get out of there, you damned cowards! I'd like to see every one of you overboard!*" The men fled, ignorant of the fact that the Second Officer had never loaded the revolver. He then turned to the others and shouted above the sound of escaping steam, "*Women and children into this boat!*"

Mrs. Minnie Coutts was traveling with her two sons William and Neville. After the collision, they became hopelessly lost below decks. A crewman directed her up to the lifeboats, but she found her way barred by a gate. Another crewman passed who gave her alternate directions as well as his lifebelt and then asked, "*There ma'dam if you're saved, pray for me.*" She found her way to the Boat Deck where Minnie and Neville got into No. 2, but Lightoller refused to let her elder son William enter because he looked too old. She was finally able to persuade the officer to let the nine-year-old pass.

Standing by No. 2 was Donald Douglas, his wife Mahala and her maid Berthe. As the boat was loaded, twice Mrs. Douglas begged her husband to come with her but he refused.

First he proclaimed, *"It would make me less than a man."* Again she pleaded and again he declined saying, *"No, I must be a gentleman,"* as he turned away from the boat, and disappeared up the deck.

Mrs. Elizabeth Robert had, in the space of seven years, buried two husbands, the second in December 1911. She and her daughter chose to run away from their sorrows by going on holiday in Europe. By the time they returned, her party had grown to include not only her daughter Georgette, but niece Elizabeth, and her maid Emilie Kreuchen. The four ladies entered the lifeboat together.

With No. 2 now ready, Wilde, seeing that the boat was short-handed, turned to the Second Officer, *"You go with her, Lightoller,"* he said. *"Not damn likely,"* Lightoller snapped back. No one could say what, if anything, was The Chief Officer's reply. An insubordinate Second Officer was, at the moment, the least of his problems. Lightoller then put Boxhall in charge of the boat. With a crew that included one Saloon Steward, one deck hand, and one assistant vegetable cook, the Fourth Officer lowered away. The cutter was designed to hold 40 people. There were 17 in the boat. The time was 1:45am.

Approximately 40 miles away, the *RMS Carpathia* was coming hard but still not hard enough to suit her Commander. Rostron called his Chief Engineer to the bridge and demanded more power. Halfway through a *"But, Sir,"* his Captain cut him off at the pass. *"Turn off the hot water, turn off the lights, turn off anything you want, but I must have more speed!"* Throughout the ship's boiler rooms, safety valves were being lashed down. The *Carpathia* took off at 14 knots, then 14.5, 15, 15.5, 16 knots. More coal meant more steam, 16.5 knots, then 17, and ultimately a stupendous 17.5 knots. The little Cunarder was going faster than her speed of design, faster than she ever had or ever would again.

Up on the bridge deck in the *Carpathia's* wireless room, Cottam got a desperate message from the *Titanic*. It would be the last time the "W/O" heard from the crippled liner.

Topside the situation continued to deteriorate. The slant of the deck was steeper and the ship's modest list to starboard had become a severe list to port. Wilde took charge, *"Everyone on the starboard side to straighten her up!"* Passengers and crew alike moved across the Boat Deck, as the ship sluggishly returned to an almost even keel.

The last two yard-built boats [No. 4 and No. 10] would be loaded simultaneously. No. 4 should have been in the water an hour ago. Captain Smith can take the blame for this one. The Captain thought passengers should be loaded from the Promenade Deck rather than the Boat Deck. It would be easier on them and clear the Boat Deck quicker. On her sister ship *Olympic* [Smith's previous command] this would be possible. However, this was the *Titanic*. The forward end of the Promenade had been enclosed to create the Parlor Suites with their own private deck. The boat was lowered to B-Deck and then raised back up to the Boat Deck. Passengers [almost all of them from First Class] who walked down to the Promenade now turned around and walked back up. Crewmen were sent down to open the windows while Lightoller left to launch No. 8 and No. 6. Now the flower of New York and Philadelphia society stood waiting in queue for one hour. These were people not used to waiting and certainly not standing in line. The comedy of errors continued.

Someone noticed the sounding spar was directly below No. 4. Two men were sent to chop it away, but they couldn't find an axe. Finally the windows were opened and the passengers were sent down to B-Deck yet again. Marian Thayer, her blue blood boiling, spoke for all,

"Just tell us where to go and we will follow. You ordered us up here and now you are sending us back!"

At 1:45am the Second Officer returned. By now the *Titanic* was listing to port, meaning that the boats would not be flush with the ship's side. There was, in fact, a three-foot gap between the liner's hull and the lifeboat. Quick thinking seamen hastily gathered together deck chairs to bridge the distance. Then with one foot in the lifeboat and the other in a windowsill, Lightoller, as usual, called for *"women and children only!"*

Madeleine Astor, in company with her maid [Rosalie Bidois] and nurse [Caroline Endres] had endured the long wait and now boarded, with the help of her husband. Mr. Astor asked Lightoller if he could join her, she being in a *"delicate condition."* He refused, telling him: *"No men are allowed in these boats until the women are loaded first."* Astor told his wife, *"The sea is calm, you'll be all right. You're in good hands. I'll meet you in the morning."* Astor then asked the boat number, to which Lightoller replied, *"Number four."* Colonel Gracie was sure Astor wanted the number to help find his bride. Lightoller was just as sure that somebody would hear about it.

Next up would be the Ryersons. The grieving family had been racing for home, in time to bury a son killed in a car crash. Arthur led his wife Emily, three children and maid Victorine Chaudanson. For Miss Chaudanson, just getting back from B-Deck had been an adventure. She had returned to the Ryerson's cabin in order to retrieve their valuables. While looking around, she was horrified to hear the door lock turn shut. A steward was securing the rooms. The servant's screams alerted him and he let her out. On deck Mr. Ryerson gave her his lifebelt, while Mrs. Ryerson led her son Jack to the window, but Lightoller stopped her right there. *"That boy can't go!"* By now the attorney was in no mood for a discussion with the Second Officer. *"Of course that boy goes with his mother...he's only thirteen!"* Lightoller gave in, but was heard to grumble, *"No more boys."* Her banker husband and son Harry helped Eleanor Widener into the boat. Marian Thayer, having properly told off the deck steward, now turned her attention to a more serious matter. There was no sign of her son Jack. She would have to enter Lifeboat No. 4 without him.

With Quartermaster Walter Perkis in charge, four crewmen to row and thirty passengers [six of them servants of First Class travelers] No. 4 was finally lowered at 1:50am. Mrs. Ryerson was stunned to see the surface of the water was less than fifteen feet below the boat.

As one by one the boiler rooms flooded and went "off-line," there was less power available throughout the ship. Phillips and Bride felt the pinch in the Wireless room. They were barely hearing some messages, and some others not at all. Worst of all, other ships were starting to have trouble hearing *Titanic*. Their *"state-of-the-art"* system was far less than that without sufficient energy.

The Marconi Company was not the only game in town. All ships and shore stations used a three-letter code to identify themselves. [*Titanic* was MGY, Cape Race was MCE, etc.] The first letter "M" meaning *"Marconi."* On German ships the first letter was "D," for *"Deutsch,"* because they had [of course] their own system, *Telefunken* [as in *"Gesellschaft für drahtlose Telegraphie System Telefunken"*].

As with all things British vs. German in this period, their relationship was not overly friendly and, at times, downright hostile. If an *"Entente Cordiale"* existed between the rival shipping companies of the two nations [and that was a massive "if"], it did not reach into the wireless rooms of their liners. Witness this exchange, between an exhausted Jack Phillips and the Norddeutscher Lloyd's *SS Frankfurt*. [Keep in mind they had virtually the same conversation at 12:35.]

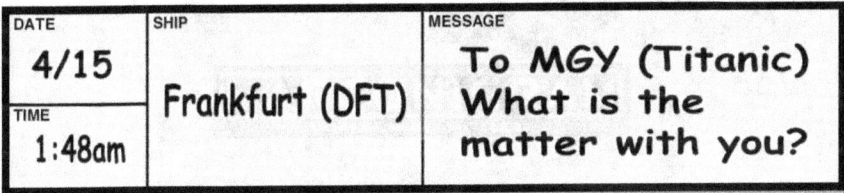

Phillips was running on fumes. All night long he had graciously responded to insipid questions from other Marconi equipped vessels, several from the *Olympic*. [*"Are you steering south to meet us?"*] Now his patience was gone. The *Telefunken* operator on the *Frankfurt* was, most assuredly, in the wrong place at the wrong time.

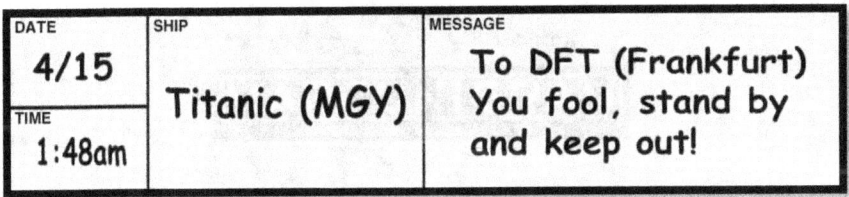

On the bridge of the *Californian,* Second Officer Stone and his apprentice Henry Gibson continued to watch the liner. They had seen her fire a total of eight rockets. *"A ship is not going to fire rockets at sea for nothing."* Something else caught Stone's eye. *"Have a look at her now. She looks very queer out of the water, her lights look queer."* Then Gibson observed, *"She looks rather to have a big side out of the water,"* and he agreed that, *"everything was not all right with her,"* that she had *"a case of some kind of distress."*

Mrs. Emily Ryerson, First Class passenger, *RMS Titanic*

Henry Tingle Wilde, Chief Officer, *RMS Titanic*

By 2:00am the ship appeared to be leaving the area. A few minutes later Gibson went down to the chart room and informed Captain Lord as such and that eight white rockets had been seen. Lord, who later said that he had been asleep [at the inquiries he would claim to have no memory of the visit], asked whether they were sure of the color. Gibson answered, *"Yes"* and went back to the bridge.

Across the water Murdoch was preparing to load the last of the yard-built boats, No. 10. As with No. 4, the ship's list required deck chairs to span the three-foot chasm between the boat and the ship's side.

Miss Gretchen Longley of Hudson, New York, climbed across with her aunts Kornelia Andrews and Mrs. Anna Hogeboom. When Gretchen boarded in Southampton, she found a farewell letter waiting in her cabin.

> *Good weather*
> *Refreshments*
> *Every desire*
> *Tommies to burn*
> *Chocolate icecream*
> *Heavenly evenings*
> *Entire meals*
> *No regrets*

The first letter of each line spelled out G-R-E-T-C-H-E-N. She had no idea who sent the note.

Canadian Real estate speculator Mark Fortune brought his wife Mary and four of their children to the boat. Mary and her three daughters climbed aboard. Mr. Fortune and his 19-year-old son Charles did not. Mrs. Ada West needed help with her daughters, four-year-old Constance and ten-month-old Barbara. Ada's situation was further complicated by the fact that she was pregnant.

The officers carried on with the loading, as the sense of urgency continued spiraling upward. The Boat Deck was just a few feet above the water line. The boarding attempt of a young French woman nearly ended in disaster when her jump into the lifeboat fell short, and she dropped into the gap. She caught the gunwale of the lifeboat while her feet found the railings on the deck below, and she was pulled back on board the ship. The young lady made it into the lifeboat safely on her second attempt. Children were rushed aboard, one baby literally being thrown in and caught by a woman passenger.

And the British Officers, as we have seen, were a little lacking when it came to tolerance. As No. 10 was being lowered a passenger whom Fifth Officer Lowe later described as a *"crazed Italian"* jumped aboard. Again, a person who didn't speak English was singled out. If anyone behaved in a manner that was either brutish or uncouth, the explanation was he must be *"some kind of foreigner,"* more often than not a *"Southern European."* Masabumi Hosono, the lone Japanese passenger aboard, took the last place in the last wood boat.

Murdoch put Able-Seaman Edward Buley in charge and lowered away. It was a very short trip indeed, as the waterline was only ten feet below the Boat Deck. There were 57 people in the boat, and Buley had just one Able Seaman, Frank Oliver. The First Officer looked at his watch. The time was 1:50am.

Engelhardt Collapsible Lifeboat

There was still power, but at the moment there was no one at the key. Phillips and Bride were busy struggling with a stoker who entered the wireless room and tried to steal Phillip's lifebelt. Bride held the man, while Phillips hit him repeatedly. The stoker finally slumped to the floor, having been knocked unconscious.

The Engelhardt collapsible boat was a marvel of simple engineering. Pull up the canvas sides, insert the slats and floorboards and *Voilà!* Instant lifeboat. [Although, there were also heavy and unwieldy.] Wilde and Murdoch went to work preparing Collapsible "C," as crewmen retrieved the boat from its stored position, quickly set it up and then attached it to the cutter's davits.

The majority of the forward boats had gone by this time and most of the crowd on deck had moved aft as *Titanic's* bow dipped deeper into the water. The "C" boat was rushed by a group of stewards and Third Class passengers who tried to climb aboard but were driven back by Chief Purser McElroy. McElroy fired two warning shots into the air, while Murdoch tried to hold back the crowd.

Two First Class passengers, Hugh Woolner and Swedish Army Lieutenant Mauritz Björnström-Steffansson [both members in good standing of *"our coterie"*] came to the officers' assistance and dragged out two stewards who had made it into the lifeboat. Another familiar gentleman was helping round up the passengers. J. Bruce Ismay had apparently survived the tongue-lashing he received from Lieutenant Lowe and now was back to lend a hand.

Mrs. Latifa Baclini boarded with her three children, the youngest being just nine months old. Fourteen-year-old Jamila Nicola-Yarred and her eleven-year-old brother Elias where traveling alone on their way to Jacksonville, Florida. They had gone back to Third Class to recover their traveling money, but the passageway was already flooded. Thankfully there was still room for the pair in "C" boat.

On the way to the Boat Deck, Mrs. Catherine Peter-Joseph lost contact with her son Michael. A distraught mother and her two-year-old daughter were jostled into the waiting collapsible.

William Carter, whose twenty-five horsepower custom built *Renault* automobile was now under water in the *Titanic's* forward hold, saw his family safely into lifeboat No. 4. He joined Harry Widener and advised him to try for a boat before they were all gone, but Harry replied, *"that he would rather take a chance and stick with the ship."* In point of fact, a boat would have been his only chance. Harry couldn't swim. After Wilde repeatedly called for women and children to enter, a number of men took up the remaining spaces in the lifeboat. As the boat was about to be lowered, Carter took one of the last two places; the final seat was taken by a gentleman named Ismay.

Captain Smith, who was watching events from the starboard bridge wing, ordered Quartermaster George Rowe to take command of the boat. By now *Titanic* was listing heavily to port, and the boat collided with the ship's hull as it descended towards the water. Those aboard frantically used their hands and oars to keep the boat clear of the side of the ship, and the protruding rivets. Collapsible "C," with room for 40, was lowered with 43 aboard. It was now 2:00am.

On the other side of the bridge, the scene was repeated as Lightoller and Moody prepared Collapsible "D."

With water flowing into the Well Deck and "D"' in all likelihood, the last boat, the situation was getting desperate. Crewmembers, with assistance from Woolner and Björnström-Steffansson and arms interlocked, were forced to form a circle around the boat to ensure that only women and children could get through. One of the first in the boat was four-year-old Michael Peter-Joseph, alone and unaware that his mother and sister were at that very moment across the deck, having just entered Collapsible "C."

Jacques Futrelle's wife Lily made her way to the Boat Deck and encountered a group of men with *"smoke-blackened faces"* standing silently in a group staring at her. She later commented that they said nothing but their eyes seemed to say, *"at least you have a chance, we have none."* An all too common scene in these last minutes was the passing of notes from the people still on deck to passengers entering the lifeboats. A final message to loved ones ashore.

"Mrs." George Thorne, the mistress of George Rosenshine, who had been traveling *incognito* under an assumed name, was helped into the boat. Her real name was Gertrude Maybelle Thorne, and while she escaped in "D," her erstwhile husband did not.

Michel Navratil, who had stolen his two sons from his estranged wife, now faced the horror of trying to escape from a doomed ship. With the help of a friend, he brought three-year-old Michel and two-year-old Edmond to the Boat Deck, and just before placing his children in the collapsible, he gave them a final message,

> *"My child, when your mother comes for you, as she surely will, tell her that I loved her dearly and still do. Tell her I expected her to follow us, so that we might all live happily together in the peace and freedom of the New World."*

About twenty people were on board the collapsible when it started down under the command of QM Arthur Bright. Woolner and Bjornstron-Steffenson, having helped load "C" and "D," jumped into the boat as it was being lowered from the promenade deck, with Woolner landed upside down and his friend half-out before being pulled in. Nearby Edward Kent did not try to save himself.

Jacques Futrelle, having seen his wife safely into a boat, now stood on the starboard bridge wing, sharing a last cigarette with John Jacob Astor. Neither the journalist or the millionaire would try to enter a boat, nor would they assist in the loading. The two men were never seen alive again. Close by [as always] stood Astor's manservant Victor Robbins. Like his master, he would make no attempt to escape. As for *"Kitty,"* she had been kept in the kennel on F-Deck. The legend says that Astor himself went down and opened the cage to free his pet. And while some First Class women managed to sneak their *"toy dogs"* into lifeboats, safe to say Lightoller would not welcome a fifty-pound Airedale to board. *"Kitty"* was last scene running up and down the Boat Deck. The time was 2:05pm. The first two Englehardt boats had been lowered from the davits, the second pair would not.

In all the months of planning and designing the *Olympic*-class liners, someone had made a huge blunder. Collapsibles "A" and "B," each with a dry weight of over 4,000 lbs. were not stored on the Boat Deck. They were on the roof of the Officer's quarters, one deck above. [This was an absurd place to position a boat.] With precious little time remaining, Murdoch and Lightoller would somehow have to get the boats down from the roof, set them up, re-raise the falls from the surface, attach the craft to the cutter's davits and then swung out. All of this before they could be loaded and lowered.

Captain Smith on the Bridge, *RMS Titanic*

Chapter 9
The Abyss

"No longer mourn for me when I am dead,
Than you shall hear the surly sullen bell.
Give warning to the world that I am fled,
From this vile world, with worms to dwell."

William Shakespeare

The bulk of *Titanic's* Third Class passengers had been kept below. At 1:30am the women were released to make their way as best they could to the Boat Deck. It wasn't until 2:00am that the men were released. They arrived on deck to find that all the lifeboats had gone. Now this mass of humanity, seemingly forgotten, moved up the deck, to the Well Deck and eventually, the Poop Deck.

In the remaining boiler rooms and engine spaces, work continued. The fans and superfluous equipment had been turned off to save power and the engineers' carefully doled out what little was left. They had made their decision, without receiving orders. Chief Engineer Joseph Bell and the senior staff would stay at their posts, no matter what. And stay they did, no one even tried to get out. The *Titanic* would pass with all of her engineers.

Andrews had done all he could, and now with his seemingly boundless energy and vitality ebbing away, he faced the inevitable. He was last seen in the First Class smoking room staring at the painting "*Approach to Plymouth Harbor*" above the fireplace, his life belt lying on a nearby table. Steward John Stewart called out to him, "*Aren't you going to have a try for it, Mr. Andrews?*" The builder of the *Titanic* did not respond. Thomas Andrews would go no further. This was as good a place as any to die.

In the wireless room, Phillips and Bride were fighting a losing battle with less and less power. While Bride adjusted the Marconi's main transmitter motor-generator field regulators in an attempt to compensate for the loss of wattage, Phillips kept sending. The First Operator continued calling "*CQD,*" only now the other ships were unable to receive his faint signals. The *Virginian* heard what they believed to be a test signal from *Titanic*, but no message.

The water was now just below the bridge and lapping the cutter davits on the Boat Deck. The *Titanic* was sinking faster than before as the Atlantic had found another point of entry. The open hatches on the well deck were allowing seawater to flood the bow.

The staff of the *á la carte* Restaurant never had a chance. Abandoned by the crew since technically they were not crew, they were left to fend for themselves. Their plight enhanced by the fact that few of them spoke English. Not one made it to a lifeboat. Signor Gatti and his staff of sixty-eight perished when the *Titanic* foundered.

On the roof of the officers' quarters, Murdoch and Lightoller continued their struggle with Collapsibles "A" and "B." The Officers rigged makeshift ramps from oars and spars, down which they slid the boats onto the Boat Deck. Mass would trump their ingenuity however. The first boat, "B" broke through the ramp and landed upside down on the Boat Deck. Boat "A" reached the deck right side up and was being attached to the falls by Murdoch and Moody when it was washed off *Titanic* at 2:15 am. In the chaos, the canvas sides were not pulled up and the boat drifted away from the ship half-submerged and dangerously overloaded. "B" would leave in the same manner, except the craft was empty and wrong side up.

As the stern slowly started to rise from the water, a banging, crashing sound was heard from inside the hull. Anything that wasn't bolted down [and some things that were] came smashing forward. Furniture, twenty-nine boilers, glassware, beds, a Parsons turbine, the ice making machine, forty tons of potatoes, a *Renault* automobile, five grand pianos, hot presses, switchboards, steamer trunks, two Third Class bath tubs, dinner tables, and more made their way towards the bow to the tune of a cacophonous din.

Phillips would not give up. He kept on trying to send another distress signal. A moment after he touched the key, the lights went out and the wireless stopped working.

DATE	SHIP	MESSAGE
4/15	Titanic (MGY)	CQ...
TIME 2:17pm		

Far below in the engine room, the master circuit breaker had blown. Engineers re-set the breaker, and it blew again. The rising tide had shorted the system. The *Titanic* was plunged into darkness.

DATE	SHIP	MESSAGE
4/15	Virginian (MGN)	To MGY (Titanic) Suggest you try emergency set.
TIME 2:17am		

On the bridge Captain Smith, realizing their was no more time, moved from station to station, releasing the men from their duties. At last decisive, he ordered his crew to abandon ship and urged them to save themselves. "E. J." then made one last visit to Bride and Phillips in the Wireless room.

> *"Men, you have done your full duty. You can do no more. Abandon your cabin. Now it's every man for himself. You look out for yourselves. I release you. That's the way of it at this kind of time...every man for himself."*

As water started to fill the Marconi room, the operators made a run for it. Bride worked his way to the collapsibles just forward, while Phillips headed aft. John Phillips, the First Wireless Operator of the *RMS Titanic*, would never be seen again.

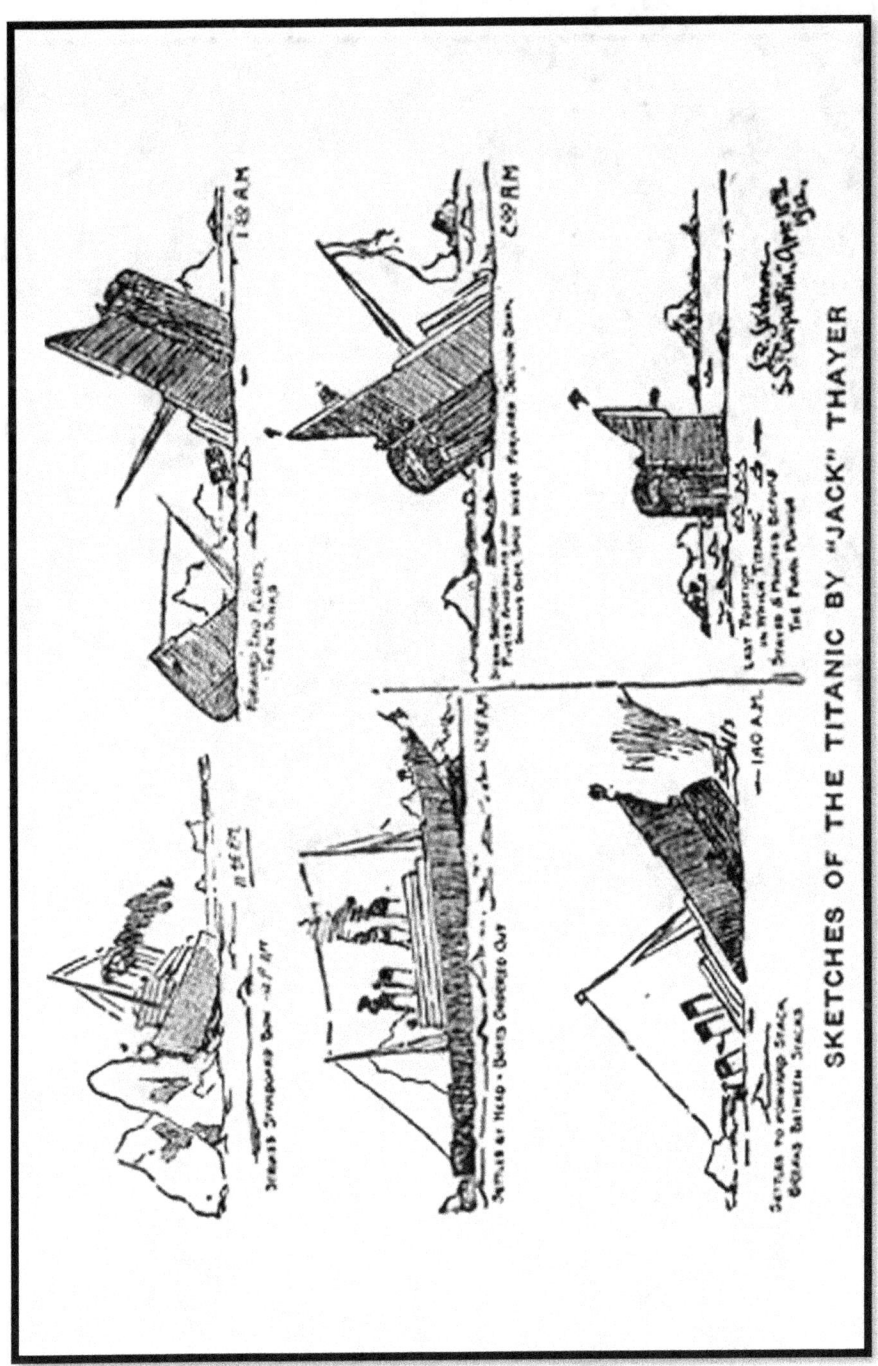

Jack Thayer's sketches, April 15, 1912

Stöwer Portrait: Sinking of *RMS Titanic*

On the roof of the First Class lounge, Wallace Hartley's musicians had stopped playing ragtime. The end was at hand, and it was time for something appropriate - but what? This depended on whom you asked. The two candidates were "*Nearer My God To Thee*," and "*Autumn*," and if it was "*Nearer*," which version? The uplifting American, or the mournful British? Survivors took both sides with Harold Bride, possibly the most credible witness, reporting that it was "*Autumn*." The majority of the passengers claimed they heard "*Nearer My God To Thee*," the hymn Wallace Hartley taught to his choir. Hartley himself may have answered the question. Former bandmates claimed he told them his last song on a dying ship would be, *"Nearer, My God, to Thee"* or "*O God, Our Help in Ages Past.*" The story persists that when the *Mackay-Bennett* recovered his corpse; in the case strapped to his waist was the sheet music for "*Nearer My God To Thee*." I have no dog in this fight. Hartley and his ensemble will be remembered, not for what they played, but where and when. Three orchestra members were washed overboard while the other five, including Hartley, went down with the ship.

> *"Many brave things were done that night, but none were more brave than those done by men playing minute after minute as the ship settled quietly lower and lower in the sea. The music they played served alike as their own immortal requiem and their right to be recalled on the scrolls of undying fame."*

Now in total darkness, every one in the lifeboats watched the *Titanic*, with one notable exception. J. Bruce Ismay sat with his back to the disaster he helped to create, never turning around. Her stern rose higher into the black night, revealing her three glistening bronze propellers. In desperation some began to jump from the ship.

Rev. Thomas Byles, Second Class passenger, *RMS Titanic*

Others chose to put their faith in God. Father Thomas Byles had passed up two chances to enter a lifeboat. Now with men and woman huddled around him, he recited the rosary, heard confessions, and gave absolution to more than a hundred who remained on the stern.

Water swept across the Boat Deck, washing the inverted collapsible, passengers, officers and crew into the sea. To his horror, Bride discovered he was now trapped beneath the overturned hull. *Titanic's* increasing angle in the water caused the stays supporting the forward funnel to snap, and it toppled into the sea, crushing swimmers beneath it and washing Collapsible "B" away from the sinking ship. Somewhere under that fallen stack was probably the body of John Jacob Astor. When his corpse was recovered by the *Mackay-Bennett*, the body was found caked with soot.

There was one final scene left to play, *"Götterdämmerung,"* the death throes of a great lady. Her stern, now completely out of the water, without any support and subject to forces it was never meant to endure, started to break away. Once again came the popping of rivets, the opening of seams, the failure of iron, and the ripping of steel, as the liner was slowly being torn in two. The bow section, with no buoyancy remaining, immediately began to plane its way toward the ocean floor. The stern section, now free of the mass of the bow, settled back for a moment. Then slowly it started to rise again. The hundreds of people clinging to railings and stanchions on the afterdeck began falling into the sea. The stern continued to an almost vertical position, paused momentarily, and then started down. In less than a minute, Captain Smith's Blue Ensign disappeared from sight. She was gone. All that remained was a calm sea, twenty lifeboats, and the anguished cries and moans of the dying. Third Officer Pitman, at the tiller of Lifeboat No. 5 noted the time. 2:20am.

| DATE 4/15 TIME 2:20am | Virginian (MGN) | To MKC (Olympic) Have you heard anything about Titanic? |

| DATE 4/15 TIME 2:20am | Olympic (MKC) | To MGN (Virginian) No. Keeping strict watch, but hear nothing more from Titanic. |

In 1688, Edward Lloyd opened a coffee house on Tower Street in London. The house soon became a favorite of both seamen and businessmen. News and information about ships changed hands and eventually groups of underwriters [known as syndicates] began selling insurance to ship owners. As the British Empire grew, so did the underwriter's business. Lloyd's, having long since outgrown the cafe, and now quartered in *The Royal Exchange,* became known around the world as *Lloyd's of London.*

In the center of the subscriber's room there hangs a ship's bell [recovered from the wreck of the frigate *HMS Lutine*]. By custom, the *Lutine Bell* was struck once for bad news and twice for good news. On the morning of April 16, 1912 the bell was struck once. The report that all of Great Britain was waiting for had reached the floor. *RMS Titanic* had passed from the British Registry. The news was both tragic and costly. *Lloyd's* held the policy on the liner.

Chapter 10
Adrift

"A small craft in an ocean is,
or should be,
a benevolent dictatorship."

Tristan Jones

In water with a temperature of 28°F, the human body has a life expectancy of anywhere from fifteen to thirty minutes. Shock is instantaneous and shortly followed by paralysis. Ironically, vigorous exercise, such as swimming, will only accelerate the process. Regardless, the end result is always the same, death by hypothermia.

For the nearly one thousand people in the water, the struggle to stay alive had begun. Some thrashed about, trying to swim, while others clung to wreckage. For those who jumped, many had already lost consciousness. And with it all came the sounds: the screams, the moans and the frantic cries for help in a dozen languages.

To those close by, their best chance came from the two collapsibles. Mrs. Rhoda Abbott had jumped from the railing with her two sons. She was able to scramble into "A," but in the process lost her boys. Harold Bride was finally able to get out from under "B." The wash from the falling funnel pushed Jack Thayer up against the same boat.

Lightoller was swimming near the foremast when he was sucked under as water flooded down one of the forward ventilators. He was pinned there against the grating for several seconds. A blast of hot air from within the depths of the ship erupted out of the ventilator and blew him to the surface. He saw "B" floating near by and swam over to it. Lieutenant Charles Herbert Lightoller, the *Titanic's* senior surviving officer and the holder of an Extra Masters Certificate, [meaning he was qualified to command the *Titanic* herself] was now in command of an upside down, water logged Englehardt collapsible.

> "How anyone that sought refuge on that upturn boat survived the night is nothing short of miraculous. Some quietly lost consciousness and slipped overboard. No one was in any condition to help."

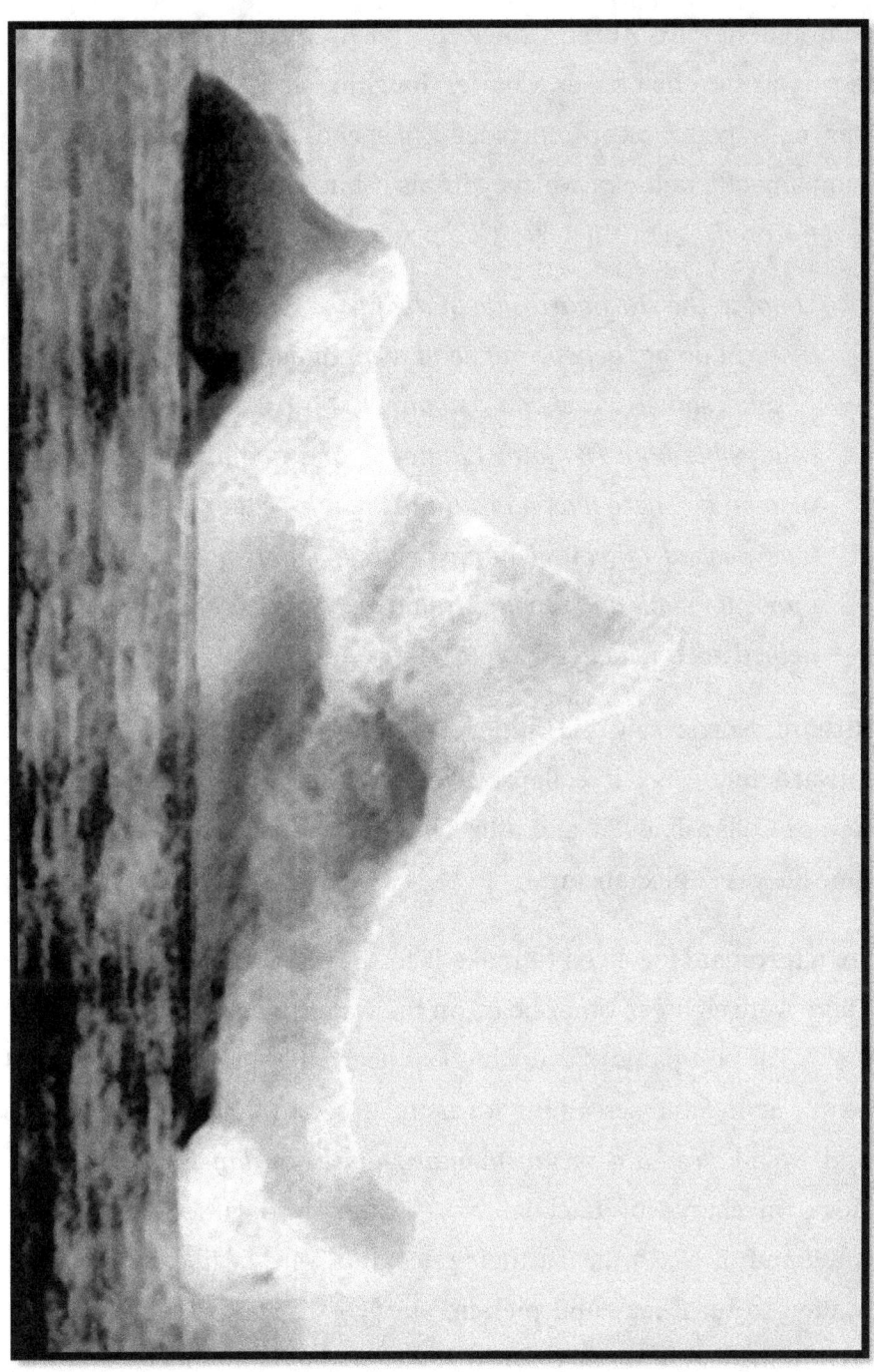

Iceberg presumed to have sunk *RMS Titanic*

Algernon Barkworth, a Justice of the Peace made it to "B," as did bedroom steward Alfred Crawford and Archibald Gracie. And then there was the chief baker, Charles Joughin. Joughin, believing there was no way to escape, decided to spend his final hours in the company of a bottle of whiskey in his cabin. By 2:15am he had been *"over served."* [Fortunately, *"God protects children and drunks."*]

> *"I got to the starboard side of the poop; found myself in the water. I do not believe my head went under at all. I thought I saw some wreckage, swam towards it and found collapsible boat (B) with Lightoller and about twenty-five men on it. There was no room for me. I tried to get on, but was pushed off, but I hung around. I got around to the opposite side and cook Maynard, who recognized me, helped me and held on to me."*

Richard Norris saw his father and many others crushed by the forward funnel as it collapsed. The resulting wave washed him toward Collapsible "A" and after clinging to the craft's side for some time he was hauled aboard.

So, a fortunate few lived to make it to the last two collapsible boats. There were eighteen other boats on the water that night. Where were they? The occupants of every lifeboat heard the sound of hundreds of people crying and screaming for help. A sound that Dorothy Gibson said would, *"remain in my memory until the day I die."* George Hogg, in charge of Lifeboat No. 7 and Third Officer Pitman, in command of No. 5, heard their cries and were of like mind. They wanted to turn back and pick-up survivors. The passengers in his boat immediately shouted down Hogg. Pitman announced that they were going back, *"Now men, we will pull toward the wreck!"*

The women protested, one begging a steward, *"Appeal to the officer not to go back. Why should we lose all our lives in a useless attempt to save others from the ship?"* Pitman gave in to their requests.

The roles were very different in Lifeboat No. 6. QM Robert Hichens had only two men to row and thought it pointless to turn back. Some of the passengers, however, pushed for the boat to pick up survivors: *"We are not going back to the boat. It's our lives now, not theirs,"* came the reply, but that was only the beginning. Captain Smith had ordered Hichens to stand by and pick up survivors. Another request to go back brought a similar response from the Quartermaster, *"There's no use going back, 'cause there's only a lot of stiffs there."* He was disobeying a direct order from the Captain. And so it went, Hichens resentful of Major Peuchen's presence, and Peuchen irked by Hichens attitude. Maggie Brown asked the QM to let the women row to help keep them warm. When he refused, she threatened to throw him overboard. He swore at her but was told to shut up by a stoker who told him, *"Don't you know you're talking to a lady?"* Mrs. Brown and Mrs. Candee now bent to the oars.

While I would never dismiss Hichens' boorish behavior, I do admit to a quandary. The lifeboats had received conflicting orders. Some were ordered to stand by; others ordered to row for the lights of the "mystery ship," while still others received no orders at all. These came from different officers, three of whom were now dead. Hichen's assessment, that his boat had just two men [plus Maggie and Helen] and could barely make headway, was in hindsight, probably correct.

Harold Lowe was in command of No. 14, and there was absolutely no doubt in his mind what to do. He gathered Boats 10, 12, 14 and Collapsible "D" together, transferred those aboard No. 14 to other boats, and with a hand-picked crew, went out looking. So, the man who, only minutes before, threatened to blow a passenger's brains out on the Boat Deck now mounted the only rescue attempt of the night. Sadly, it was much too little, far too late. The Fifth Officer rescued four survivors, one of who would be dead within the hour.

In No. 2, Boxhall suggested to the occupants that they should go back to pick survivors up from the water. They refused outright. Boxhall found this puzzling as only a short time before the women had pleaded with Smith for their husbands to be allowed to accompany them, yet now they did not want to go back and try to save their lives. To seven-year-old Eva Hart it was all a terrible dream.

> "I can remember the screams. It seemed as if once everybody had gone, drowned, finished, the whole world was standing still. There was nothing, just this deathly, terrible silence in the dark night with the stars overhead."

Fireman Frederick Barrett and the passengers in No. 13 gave no thought to going back, they just kept rowing. Eventually, when the muffled cries and moans of the dying subsided and salvation was at hand, Dodge, Beesley and the others would turn to song:

> "Pull for the shore, sailor, pull for the shore!
> Heed not the rolling waves, but bend to the oar
> Safe in the lifeboat, sailor, cling to self no more!
> Leave the poor old stranded wreck, and pull for the shore."

The women in the boats did whatever they could do to keep warm, some even manning the oars. In No. 8 the ladies did the bulk of the rowing. For most of the night, the Countess of Rothes was at the tiller. Seaman Jones, the man in charge of the boat, explained why:

> "She had a lot to say, so I put her to steering the boat. When I saw the way she was carrying herself and heard the quiet determined way she spoke to others, I knew she was more of a man than any we had on board."

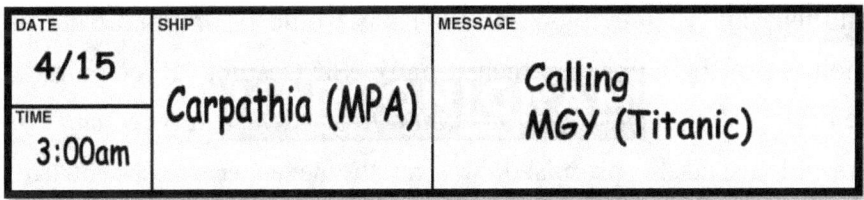

The passengers on the *Carpathia* were asked to stay in their cabins to avoid the scene on deck. It was just as well; their ship was now playing "dodge ball" with icebergs. Rostron kept missing them and refused to slow down. She was approaching *Titanic's* position and he kept pouring on the coal. One seaman remarked, *"It looked like everyman on the ship had found a shovel."* And now, finally they knew why. An hour before the stewards had been mustered in the main dining salon to hear from the Chief Steward. He told them about the *Titanic*, about their duties and then finished with a summation that Lord Nelson himself would have been proud of:

> "Every man to his post and let him do his full duty like a true Englishman. If the situation calls for it, let us add another glorious page to British history."

[He was paraphrasing, using a line that every British schoolboy could recite by heart. Nelson's exhortation to the fleet before Trafalgar: "England expects that every man will do his duty."]

DATE	SHIP	MESSAGE
4/15	La Provence (MLP)	To MLC (Celtic) Nobody has heard the Titanic for about 2 hours.
TIME 3:28am		

The No. 1 boat had room for another 30 people, but as with most of the others, its occupants showed no willingness to return to rescue those in the water. It all started to go downhill for the Duff-Gordon's when a fireman named Robert Pusey told them that the crew had lost all their "kit" [belongings] and their pay would be stopped from the moment of the sinking [which it had]. Sir Cosmo, rather irritated, retorted, "Very well, I will give you a fiver each to start a new kit!" He did just as he promised, writing the seven crewmen aboard a check for £5 each. Equally insensitive was his wife Lucy, who in the face of the greatest shipwreck in memory, remarked to her secretary, "Miss Francatelli, there is your beautiful nightdress gone."

One was inverted, another unseaworthy, three more were filled to, or over capacity. Of the fifteen remaining boats, only one made any attempt to affect a rescue. Lt. Lowe would add three to the list of the saved. Clinging to Collapsible "B," Jack Thayer didn't understand.

> "The partly filled lifeboat standing by 100 yards away never came back. Why on Earth they never came back is a mystery. How could any human being fail to heed those cries?"

In any lifeboat you were cold. If you were in No. 7, No. 11 or Collapsibles "A" or "B." you were something else, anywhere from ankle to waist deep in the Atlantic. The night continued with some rowing, some praying, some shivering, but all longing to see the dawn. Then in the distance, a passenger saw a flash of light rising in the night sky. Out in the blackness someone was firing rockets.

Chapter 11
Deliverance

*Survival is not so much
about the body,
but rather it is about the
triumph of the human spirit.*

Danita Vance

The long awaited dawn was finally breaking over the North Atlantic as the *Carpathia* reached Boxhall's position. Still running at flank speed, she had been firing her own rockets every fifteen minutes, hoping against hope that the *Titanic* could see her coming.

In the gathering light Rostron ordered the engines stopped. His heart sank as he viewed the scene and saw no sign of the great liner. If his heart was low, his pulse rate was probably quite high. As he scanned the horizon he saw that *"innumerable pyramids of ice"* surrounded his little ship. Rostron, a pious man, knew who was responsible. Of the risk taken by running through dense ice at high speed in the dead of night, he is reported to have said,

> *"I saw the ice and could only think that some other hand than mine was on that helm during the night."*

Only now, when they needed it most, Divine Providence was nowhere to be found. The calm sea, which had been a Godsend during the night, was no more. The morning breeze was freshening, and with it came waves and chop. The small craft could add a rolling sea to their list of headaches. Other patches of white now appeared on the horizon. "*Growlers?*" No, lifeboats. The craft had seen *Carpathia's* signals in the dark and had been pulling for the rockets. Rostron started his engines *"all ahead slow."* The little boats, a handful at the best of times, would be unable to maneuver effectively in the swell. [There is a big difference between *"sea worthy"* and *"sea friendly."*] The *Carpathia* would maneuver for them. The Captain positioned her between the boats and the wind, helping to shelter them to leeward. His ship had just raced flat out fifty-eight miles through a treacherous ice field in total darkness in less than four hours. Against all odds, Rostron and his steamer had made the rendezvous.

Close aboard was Boxhall with No. 2. The cutter was at *Carpathia's* gangway at 4:10am. The Fourth Officer went straight to the bridge, where Rostron posed the question; fearful of the answer he would receive. *"Titanic has gone down?"* Boxhall, shivering in the cold and with his voice breaking replied, *"Yes, she went down at about 2:30."*

Less than five miles away, there was a flat, unbroken ice field that stretched as far as the eye could see. Rostron had no time to worry about ice flows; there was a flotilla of little boats that needed his help.

Worst off were the collapsibles. The air beneath the upturned "B" boat was slowly leaking out, and the survivors were standing, trying to keep her from capsizing. Someone spotted the cluster of boats that Lowe had strung together. One single blast from Lightoller's whistle [an officer's signal] sent No. 4 and No. 12 over to her. "B" was in such a precarious state that the mere wash of another boat almost tossed everyone into the sea. One by one they were transferred, Jack Thayer [not spotting his mother, who was ten feet away in No. 4] Colonel Gracie, and lastly, the Second Officer. Among the saved was baker Joughin, whose liquid insulation had seen him through two hours of treading water in the freezing Atlantic. The *RMS Carpathia* would rescue seven hundred, and a bottle of Scotch would save one more.

The salvation of collapsible "D" would come, from all things, a sailboat. The 16 yard-built boats came with a mast, a spar and a sail. They were a nuisance and an encumbrance to all. All save one. [Some crews left the rigging on deck; others threw it over the side.] The Fifth Officer was a boatman as well as a seaman. At the first sign of a breeze, the crew of No. 14 hoisted her sail and she took off at a brisk four knots. Lowe spotted "D" upwind, low in the water. He tacked and jibed his boat over to her and took the collapsible in tow.

Lifeboat No. 14 & Collapsible "D," *RMS Titanic*

Englehardt Collapsible, RMS Titanic

Titanic lifeboat alongside *RMS Carpathia*

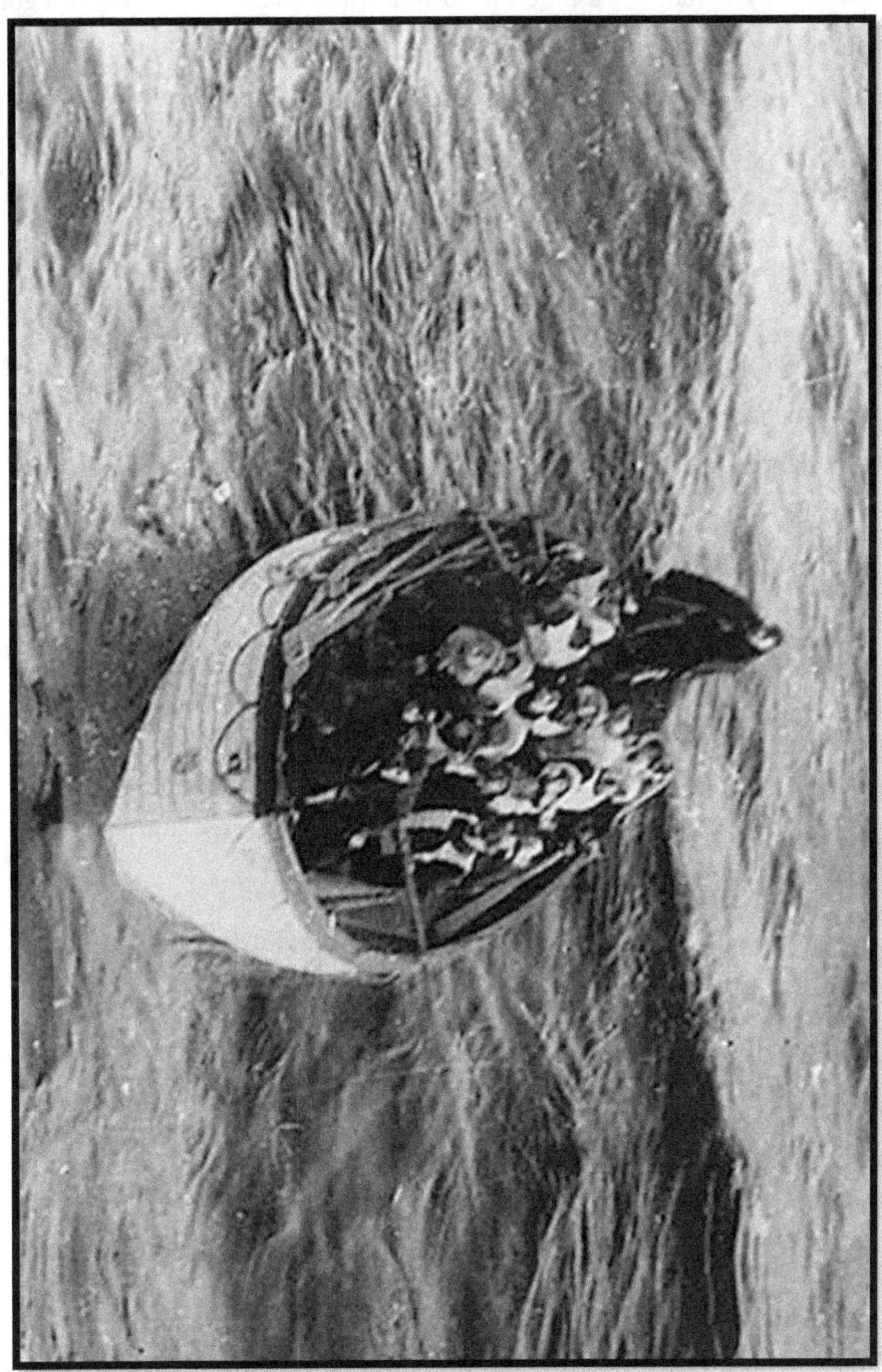

Lifeboat No. 12, *RMS Titanic*

Worst of all was "A," a mile-and-a-half away. The crew never did get her sides up, and now she lay nearly flush with the water. Someone in No. 14 spotted her. Now running before the wind, Lowe closed the distance, pulled up along side and transferred all of the survivors to his craft. Thirty people were aboard "A" when she washed off the Boat Deck, now four hours later only a dozen men and one woman, Mrs. Rhoda Abbott were alive. Hypothermia had claimed the rest.

One by one they moved to the *Carpathia*. Collapsible "C" arrived at 5:45am carrying among others, Joseph Bruce Ismay. The Line's Managing Director came aboard mumbling, *"I'm Ismay, I'm Ismay."* He was taken straight to Dr. McGhee's cabin, oddly demanding to be fed. While his passengers slept on dining room tables, Ismay insisted on a private cabin. It was there he spent the next four days. Many believed he was under the influence of opiates.

The No. 3 Boat was made fast at 6:00am. Off came the Harper entourage, Mr. and Mrs., his dragoman, and her Pekingese. Those without the strength to climb the gangway came up in a rope sling. The "singers" of No. 13 reached the ladder at 6:30am.

Lifeboat No. 14, still under canvas, came up at 7:00am carrying the survivors of "A," and with "D" in tow. The passengers were immediately rushed to the dining room for hot food and stimulants. Lowe, like any true boatman, had one more job to do. He lowered the spar, unshipped the mast, stowed the sail, and only then climbed out. No one else had done so much or so well as Harold Godfrey Lowe and the No. 14. The little boat and her skipper had performed brilliantly.

And still they came, with those now aboard lining the rail, hoping to get a glimpse of a friend or family member. Happy reunions were mixed with sudden shocks that a loved one or shipmate was missing.

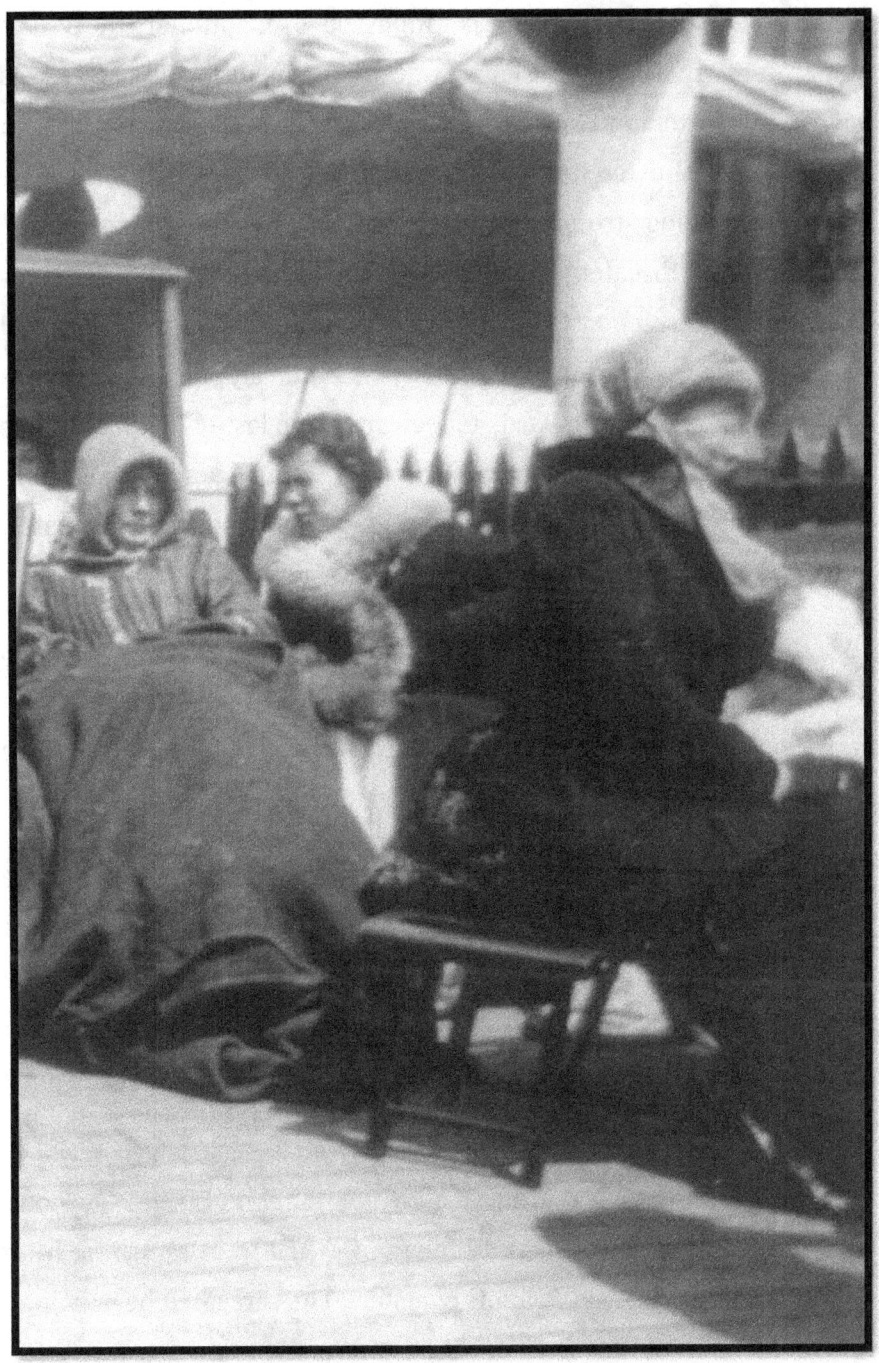

Titanic survivors aboard *RMS Carpathia*

The end was not quite in sight. By 8:15am they were all aboard except Lifeboat No. 12. Now commanded by Lightoller, the vessel was crammed with 75 passengers, 10 above her capacity. She was low in the water and could barely make way. The nearly frozen Second Officer nursed her to within 200 yards of the *Carpathia*. The wind was blowing stronger now and the waves were lapping against the gunwales, threatening to swamp the boat. Yet again it was Rostron to the rescue. With deft manipulation of the propeller and the rudder he re-positioned his liner. When he stopped her engines, No. 12 was less than 100 yards to leeward. The last of *Titanic's* boats had been off-loaded by 8:30am. The crew then began hoisting them aboard. Thirteen of the yard-built boats were pulled from the water, while the other three and the collapsibles cast adrift.

Over the horizon the German liner *Prinz Adalbert* was working its way through the ice, unaware of the *Titanic*. Those passengers up early on the promenade deck were astonished to see close aboard, an iceberg with a scar of red and black paint along it's waterline.

The passengers and crew of the *Carpathia* did all they could for the survivors, with food, coffee, blankets and dry clothes. The mood of the saved ranged from grateful to sullen. Most were happy just to be alive, while others were not. A steward bringing food to a group of women huddled forward was sent on his way when one of them spoke for all, *"Go away, we've just seen out husbands drown!"*

Further down the deck there was a joyous reunion. Sisters Charlotte Appleton, Caroline Brown, and Malvina Cornell were returning from England and the funeral of another sister. The three had been helped into boats by Colonel Gracie, and now were stunned to see their uncle, *Carpathia* passenger Charles Marshall rushing to greet them.

The remaining members of *"our coterie"* gathered for the final time. Of the seven writers only Mrs. Candee, along with Woolner, Björnström-Steffansson and Col. Archibald Gracie, were safe. But Gracie's time in the freezing water had taken its toll. The aftereffects of shock and exposure meant the Colonel would not live to see 1913.

As the boats were being loaded aboard the *Carpathia,* the Captain plotted his next move. His ship really wasn't designed to handle seven hundred guests who had just "dropped in." Limited provisions demanded a return to New York, and so he plotted the course. Cottam in the Wireless room now heard from the *Olympic.* She offered to take on the passengers. Rostron was appalled. He couldn't conceive of anything worse that subjecting the survivors to another trip in a lifeboat to a ship identical to the *Titanic.*

I bow to no man in my respect for Arthur Rostron. However, this one time I believe even the Captain had an agenda. It was his ship that answered *Titanic's* call for help, his ship that raced through the night, risking everything to reach the survivors. His ship that took them all aboard, and it would be his ship that would deliver them safely to the North River piers in Manhattan. There would be no sharing of glory.

At 9:15am he was called back to the bridge. A ship was coming up from the southwest. He had seen the *Mount Temple* and asked her to continue searching for survivors. No other ship was due it the area. The *Carpathia* used semaphore flags to tell the newcomer that she had picked up all the survivors from the *Titanic* and was now inbound for New York. The Cunarder then put about and left the scene. The new ship continued the search for a time but then she also turned west. As a result of the ice, her Captain was well behind schedule. His ship, the *Californian,* was due in Boston in three days.

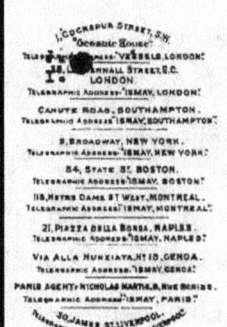

White Star (Liverpool) seeking information, *RMS Titanic*

Chapter 12
An Ocean Apart

"This blessed plot, this earth, this realm, this England"

William Shakespeare

For everyone saved, two were gone. Astor, Guggenheim, Archie Butt, the Wideners, father and son, were gone. Anders Andersson, his wife and five children, all gone. Wallace Hartley and his musicians were gone. Smith, Wilde, Murdoch, McElroy and Moody - they had done their duty well and true and now they were all gone. None of the eleven members of the Sage family would see Florida; someone else would grow the pecans. Joseph Bell and his Engineers, all gone. Pietro Gatti and the Restaurant staff, scores of stokers, trimmers, stewards, waiters and deck hands were all gone. *"Dai"* Bowen, gone. Edward Kent [with Helen's cameo] and Thomas Andrews, gone.

In the world of 1912 there was, of course, no television, and the closest thing to radio was Marconi's wireless. In this world, news, weather, sports and most certainly commentary came in a folded piece of newsprint, more commonly referred to as *"the paper."* Every big city had newspapers, usually more than one. New York had no fewer than twenty "dailies," none more prominent than the others. As with so many other things, this too was about to change.

The late shift dragged on in the newsroom of *The Associated Press* and, across town, on 42nd Street, at *The New York Times*. Feet up on the *AP* city desk, an editor named Charles Crane read an H.G. Wells novel to while away a very quiet, very news-free Sunday night.

At *the Times,* managing editor Carr Van Anda had returned from a late supper to the office. The closest thing to a lead story, the latest chapter in the on-going war between Taft and Roosevelt, was being readied for the front page. [The day before *"Teddy"* had routed the incumbent in the Pennsylvania primary.] A copy boy dozed. No one had a clue that a thousand miles away, the *"story of the century"* was breaking. The bulletin dropped from the sky, literally.

Carr Van Anda, Managing Editor, *The New York Times*

The City Room, *The New York Times*

The basket came crashing down the dumbwaiter from the wireless station on the roof. Dumbwaiter bulletins would arrive from Cape Race via Halifax [and not from *"Wanamaker's"* Department Store]. The now wide-awake copy boy grabbed it and raced to the Managing Editor. Van Anda read it once and then read it again.

The man they called "V. A." set to work. He placed a call to the White Star office in Lower Manhattan, followed by calls to *The Times* correspondents in Montreal and Halifax. At this point little was known. A half a dozen ships including her sister, the *Olympic,* had received the signals, changed course and were heading for the area. Cape Race was monitoring all traffic from *Titanic,* but now there was nothing to monitor. The stricken liner had stopped transmitting.

Van Anda, the quintessential newsman, realized the biggest story of his career had just dropped in his lap. He reshaped the front page, giving *"Teddie"* and *"Big Bill"* the bum's rush to the inside pages. There would be nothing but *Titanic* on his front page. As he worked a feeling of dread came over him. No one had heard from her for over an hour, and the last messages said that women and children were being put in lifeboats, and the engine room was flooding. With a deadline to meet, he played it straight for the morning edition.

NEW LINER TITANIC HITS AN ICEBERG
SINKING BY THE BOW AT MIDNIGHT
WOMEN PUT OFF IN LIFEBOATS
LAST WIRELESS AT 12:27AM BLURRED

The New York Times, Morning Edition, April 16, 1912

White Star Line Office, 9 Broadway, New York

In any other business it would be called the research department, but in the "paper biz," it was known as "the morgue," the final resting place for old newspapers. VanAnda had a hunch, and as any gambler will tell you, when you get a hunch, you bet a hunch. He sent the paper's staffers into the files to search for the answer to a question: *"Why would a ship in trouble stop using it's wireless?"* The answer came back. *"One reason only, the ship must have sunk."* Without corroboration he sent the city edition to press with that headline, knowing full well that a mistake could ruin him and the paper. There was no mistake. In one stroke he had transformed just another New York daily into the pre-eminent newspaper in the Western World, *"The Paper of Record."* The Managing Editor was at his desk as the city edition hit the street. Carr Van Anda had just scooped the world.

Now, suddenly, the most popular address in Manhattan was No. 9 Broadway, the New York headquarters of the White Star Line. By 8:00am newsmen filled the building. Vice President Franklin, as any good *apparatchik* would, dismissed *The Time's* story.

> "We place absolute confidence in the Titanic. We believe that the boat is unsinkable."

[A 50,000-ton ocean liner is a *"boat?"* Can you say *"landlubber?"*] Franklin failed to mention that his office was, at that very moment, frantically wiring Captain Smith and his *"boat."*

DATE	SHIP	MESSAGE
4/15	**White Star**	To MGY (Titanic) Anxiously await information and probable disposition of passengers.
TIME 8:15am	**(New York)**	

The White Star *"spin"* machine went into overdrive. Within two hours, friends, relatives and retainers were flooding the office.

The New York Evening Sun, April 15, 1912

Latest available news, *New York Sun*

Rich or poor, whether you were the son of J. J. Astor, the sister of an Irish immigrant or a trusted servant, you got the same smile and the same reassuring words. And with them came the same lies. *"The Titanic was unsinkable!" "She could float for days!" "There were enough lifeboats for everybody!"* That seemed to mollify the visitors.

Further reassurance would come in the form of one of the most monstrous hoaxes of the 20th century. Somewhere between New York and Halifax an excited amateur *"W/O"* had picked up fragments of two unrelated messages, and combined them. The messages, *"Is the Titanic safe?"* and *"Asian 300 miles west of Titanic and towing oil tanker to Halifax,"* became *"Titanic safe-towing to Halifax."* Crane and the Associated Press pronounced the message genuine. Some on the street and in the press, demanded that *The Times* print a retraction, but Van Anda wouldn't flinch. There would be no retraction. Nonetheless, a jittery world relaxed, especially at No. 18 Broad Street. The New York Stock Exchange watched IMM stock plummet in the morning, but start to rebound in the afternoon. Meantime, shares in *The Marconi Company* had gone through the roof. The "Big Board" players were convinced, the planet would continue to spin on its axis and life would go on. She was safe.

That was all well and good, but still no one had actually heard from the *Titanic*. Now those same amateur operators started to hear things, things that were not meant for their ears. Through back channel sources, Cunard got the word that their rival had lost its newest ship. Franklin heard those same reports but chose not to believe them. What in future generations would become known as "stonewalling" was now being practiced at No. 9 Broadway. A still confident V.P. had his world smashed to pieces at 6:15pm. Nightfall had brought the *Olympic* into range and she was transmitting:

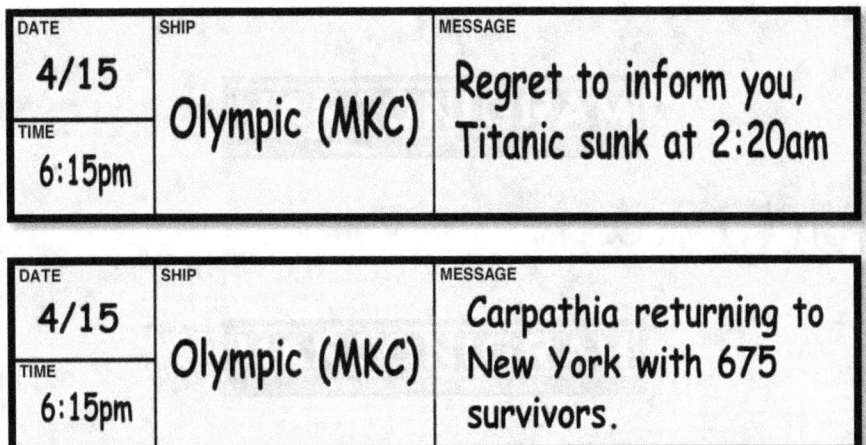

These messages had been delayed in transit. Some thought this suspicious. Were "fat cats" holding up the news so they could dump their shares of IMM? Not likely, since no proof of this has ever surfaced. Regardless, the secret did not last very long. Sharp-eyed reporters could see it in Franklin's face. He read his statement and then took their questions. These were Manhattan beat writers who could badger with the best of them. Slowly but surely, the members of the press began breaking him down.

> 7:30pm – "Gentlemen, I regret to say that the Titanic sunk this morning at 2:20am.

> 8:00pm – "Messages neglected to say that all the crew had been saved."

> 8:15pm – "Probably a number of lives have been lost."

> 8:45pm – "We very much fear there has been a great loss of life."

> 9:00pm – "It was a terrible loss of life...they could replace the ship, but never the human lives."

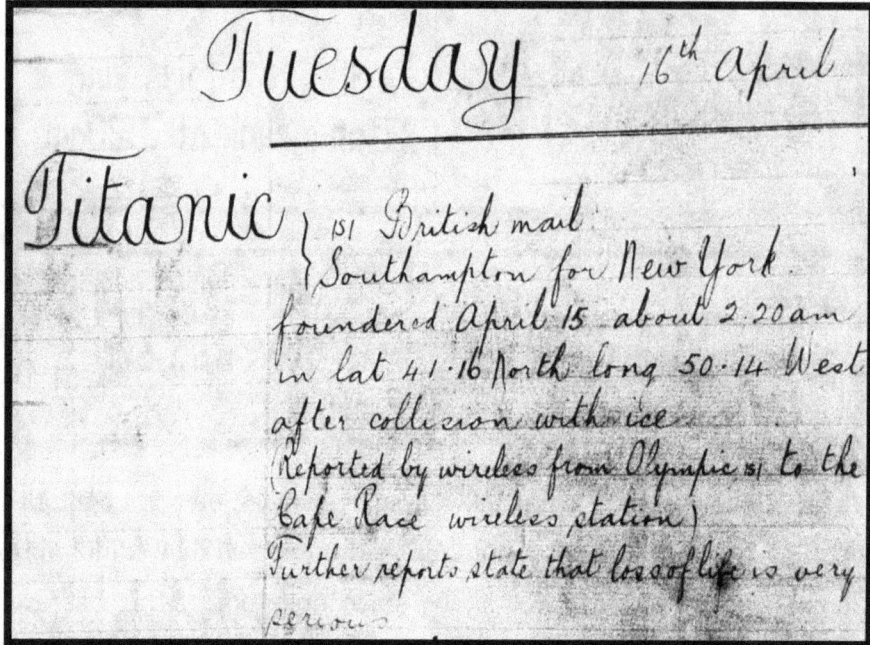

Lloyd's of London Ship Registry, April 16, 1912

Bulletin board outside *New York American*, April 16. 1912

Now the parade of friends and family would begin again. This morning they had all left smiling, now most left weeping. Vincent Astor, Florette Guggenheim and an army of relatives from Second and Third Class passengers came and went.

When the bogus news report came about the *Titanic* being towed, White Star chartered a train to take relatives to Halifax. On the special was, among others, Mrs. Alfred Hess, the daughter of Isidor and Ida Straus. Sara Hess, being the only woman on board, was basking in all the attention she was getting from the reporters. Somewhere in Maine, the train stopped, and then began backing up, all the way to Boston without a stop. A message reached the train, *"Plans have changed, Titanic's people going straight to New York."* Now the only way back to Manhattan was on the sleeper from South Station to Grand Central. Sara arrived the next morning and was met by her brother. He struggled for the words, *"Things look pretty bad."*

So, the truth was out, but since Rostron had limited wireless transmissions to lists of survivors and personal messages, no information was coming from the *Carpathia*. It was on the cover of every newspaper in the United States [and Canada and Great Britian] and like journalists everywhere, when you have no facts, you speculate. *"How could an iceberg sink a ship?" "How could a ship not avoid something so big and obvious?"* No one could get answers, not even the President of the Unites States. Taft sent a cruiser out to contact the *Carpathia* and learn the fate of his friend Archie Butt. The *USS Chester's* luck was no better than anyone else's.

The newspapers covering a story that was a half an ocean away, with precious new news [and miles behind *The Times*] decided that, *"when all else fails, pout."* The *World* pitched a fit:

CARPATHIA LETS NO SECRETS
OF THE TITANIC'S LOSS
ESCAPE BY WIRELESS

The *Evening Mail* just whined on behalf of its readers:

WATCHERS ANGERED BY
CARPATHIA'S SILENCE

Thursday, April 18, 1912 was a miserable day in New York, cold and rainy with thunderstorms lingering about. It was evening by the time the *Carpathia* passed *Sandy Hook,* reached "the Narrows," and cleared Quarantine. It is not hard to imagine the emotions of the survivors as they passed beneath Frédéric Bartholdi's Statue of *"Liberty Enlightening the World."* Rostron had one more gift for the immigrants he had saved. He held course and steamed right past the dreaded Ellis Island. Lower Manhattan and the area around the Chelsea Piers were jammed. In procession with a tugboat full of photographers, the Captain brought his ship first to Pier 59 to off load the *Titanic's* lifeboats, and then at 9:30pm she was finally warped in along Cunard's Pier 54. Many on the pierside wept openly. The *Carpathia's* passengers disembarked, and then a woman in borrowed clothes appeared at the gangway. One reporter shouted, *"Titanic?"* Softly she replied, *"Yes."* The feeding frenzy had begun.

Across the Atlantic the shock was, if possible, even more acute. The same mix of pain and grief swept *"this scepter'd isle."* The London morning newspapers heard it first and led the way. The earliest news coverage was in the Capital's papers on the afternoon of Monday, April 15th. Editors knew that the *Titanic* had hit an iceberg, but no one in England or America had yet to learn what had happened. It seemed unbelievable to the men of Fleet Street that the greatest liner in the world had sunk on her first sailing with over a thousand dead.

Arrival *Titanic* survivors, Pier 54, April 18, 1912

TITANIC SUNK.

NO LIVES LOST.

COLLISION WITH AN ICEBERG.

LARGEST SHIP IN THE WORLD.

2,358 LIVES IN PERIL.

RUSH OF LINERS TO THE RESCUE.

ALL PASSENGERS TAKEN OFF.

The White Star liner Titanic, the largest ship in the world, on her maiden voyage with 2,358 passengers and crew, collided with an iceberg in the Atlantic at 10.25 p.m. on Sunday (about 2 a.m. yesterday morning by English time).

She sank at 2.20 in the afternoon (7.20 English time).

No lives were lost.

OFFICIAL STATEMENT.

LATEST NEWS OF THE PASSENGERS.

Late last evening the White Star Line at Liverpool announced:

"Captain Haddock, of the Olympic, wires to our New York office that the Parisian reports the Carpathia to be in attendance on the Titanic. The Carpathia has picked up twenty boats full of passengers. The Baltic is turning back to give assistance. It is also reported that all the passengers are saved and the Virginian is towing the Titanic towards Halifax."

MONTREAL, Monday Evening.

It is now confirmed here, that the passengers of the Titanic have been safely transhipped to the Allan liner Parisian and the Cunarder Carpathia.—Exchange.

NEW YORK, Monday (4.30 p.m.).

According to reports published here the transfer of the Titanic's passengers was made safely in calm weather. The Baltic reported at three o'clock that she was hurrying to get to these vessels for the purpose of taking over the Titanic's passengers from them. She performed a like service for the Republic's passengers in January 1909.

Most of the Titanic's crew remained on board her. The Titanic's passengers should reach Halifax to-morrow, whence they will take train for New York.

A message from Halifax states that the Government Marine Agency has received a wireless message to the effect that the Titanic is sinking.

An official message received here via the cable ship Minia, off Cape Race, says that steamers are towing the Titanic and endeavouring to get her into the shoal-water near Cape Race for the purpose of beaching her.

A message from Montreal says that the Virginian has left the Titanic and is proceeding to Liverpool.

The Canadian Government has ordered the steamers Mackay Bennett, Lady Laurier, Rosalind, and Seal to hold themselves in readiness to go to the Titanic if they are needed. The White Star manager, Mr. Mitchell, has left for Halifax to take charge of the passengers when they arrive there.—Reuter.

LUXURIOUS LINER.

VERANDAH CAFE, TURKISH BATHS, AND RACKETS COURT.

The equipment of the Titanic is on a scale of luxury and beauty, worthy of the largest liner. The rich decorative effects introduced in the Olympic are repeated in her more recent sister ship, and the latter has a number of new features.

Among these is a Café Parisien on the promenade deck, arranged in connection with the restaurant. It is decorated in French trellis-work with ivy creepers, giving the appearance of a picturesque verandah. There is also a reception-room decorated in the Georgian style in connection with the restaurant, and the parlour suite rooms have two private promenade decks with half-timbered walls of Elizabethan design.

Over 550 passengers can dine together in the first-class dining saloon. The restaurant is Louis XVI. in design, panelled from floor to ceiling in beautifully marked French walnut, with large bay windows draped with silk curtains. This room is furnished with small tables and will seat 140 guests. The smoke-room is panelled in finest mahogany and has a verandah with green trellis and climbing plants.

The cooling-room in connection with the Turkish baths is one of the most remarkable rooms in the ship. The walls are tiled in large panels of blue and green, the dado and doors are in a rich-coloured teak, and the stanchions are also cased in teak, carved all over with an intricate Moorish pattern. From the ceiling hang bronze Arab lamps.

In the reading and writing room a great bow window at one side gives a wonderful view of sea and sky. Among other features of the Titanic's equipment are a gymnasium, a swimming bath, and a rackets court. The second-class dining-saloon will seat over 400.

The best suite of cabins costs £870 for the voyage; it has its own private promenade deck.

In all there are ten decks. The bridge deck is 550 feet long, and the promenade and boat decks over 500 feet. The bridge deck promenade is entirely enclosed in a length of 400 feet with a solid side-screen fitted with large square windows, which can be opened and shut at will.

The Titanic is the largest ocean liner at present constructed. Her dimensions are as follows:

Gross tonnage	46,328 tons.
Displacement	60,000 tons.
Length overall	883 feet.
Breadth	92.5 feet.
Speed	22.5 knots.

The height from her keel to the navigating bridge is 104 feet. The rudder alone weighs 101 tons, and is 78 feet long, while her bower anchor weighs 15½ tons.

The Daily Mail, April 15, 1912

The Easton Express, April 19, 1912

Oceanic House, White Star Offices, London

Ned Parfett selling newspapers outside Oceanic House

Editors, with very little information and almost no facts, went with the *"no news is good news"* theory. Throw in a patriotic storyline of heroic English seamanship saving the day, and you have a front page. The evening newspapers had headlines promising NO LIVES LOST. A parish church in Whitechapel held a service of thanksgiving at the news that its vicar and his wife had been saved. In fact, the couple had selflessly refused places in the lifeboats and had perished.

It was only on Tuesday that the scale of the disaster became clear. Hopes that had been raised hours before were now cruelly dashed. Coverage at first focused on the multi-millionaires and the revelation that their great fortunes could not save them from a violent death.

There was another parade of friends and relatives in London, much the same as the one in New York, only the address had changed. No. 9 Broadway had given way to No. 34 Leadenhall Street just off Trafalgar Square. Coverage of survivors would wane over time. The tabloids still couldn't get enough, but the British are a seafaring people, less interested in whom, more interested in how and why.

There was sorrow everywhere: in Belfast, the cradle of the great liner. In Liverpool, still the headquarters of the White Star Line, but most of all in Southampton. The men who passed through the Solent and by the Isle of Wight had, since Roman times, *"gone down to the sea in ships."* Now came their day of reckoning. No town anywhere in England was hit so hard. It seemed that the *"Angel of Death"* had bypassed Egypt and made its way straight to Southeast England.

The little city on the English Channel was devastated. Eighty per cent of the crew hailed from the *Titanic's* new homeport. Three quarters of them had gone down with the ship, a proportion far higher than First, Second, or even Third Class passengers.

The Sphere, April 22. 1912

The Sphere, April 20. 1912

Entire streets were hung in the British tradition with black crepe. Whole rows of houses went into mourning. As for the White Star's office in Canute Chambers, now it was their turn to deal with the crowd. Outside waited the women; the crowd was almost entirely female. Young women came with their children; older women with relatives or friends. They stood and watched as the names were posted. Some left smiling; the rest left sobbing. Many would have to walk home alone. The *London Daily Mail* described the scene:

> *"Later in the afternoon hope died out. The waiting crowds thinned and silent men and women sought their homes. In the humbler homes of Southampton there is scarcely a family who has not lost a relative or friend. Children returning from school appreciated something of the tragedy, and woeful little faces were turned to the darkened, fatherless homes."*

There were losses on Cable Street, Union Street, on every street in the town. The coal strike had devastated many a city, none more so than the Channel ports. Unemployment went sky high, and families were pawning everything they owned just to buy food. So, when jobs on the *Titanic* became available, men clamored for them. A few, the "lucky" ones, got a berth. What had seemed like a gift from heaven when their husband or brother secured a position on the new liner now revealed itself as a cataclysm. Another *Daily Mail* reporter, not quite so gifted as the first, chimed in:

> *"Many women who wait for hour after hour outside the White Star offices pathetically clung to the hope that their men being in the four- to-eight watch would have escaped in one of the boats.*

Outside the White Star Line office, Southampton, England

Titanic **crewmen returning, Southampton, England**

> *The twelve-to-four watch was the deathwatch. One drooping woman was leaning on a bassinet, containing two chubby babies, while a tiny mite held her hand. 'What are we waiting for, Mummy?' 'Why are we waiting such a long time?' asked the tired child. 'We are waiting for news of your father, dear,' came the choked answer, as the mother turned away her head to hide her tears."*

The twelve-to-four was indeed the deathwatch. These were the engineers, stokers, greasers and trimmers, who continued to work the boiler rooms as the ship went down, who stayed at their post until all hope was gone. What the women didn't know was that Ismay and White Star had made no provision for their loved ones' safety.

For all crewmen, dead and alive, their wages ceased the moment Captain Smith's Blue Ensign disappeared from sight. The British are also a charitable people. Without being asked, the sorrowful nation immediately started collecting money for a seaman's relief fund. Contributions poured in from all parts of the British Empire, and eventually reached well over two million dollars.

In America, the focus was on passengers, especially First Class passengers, and then, with whatever grief and pity remained, those in Third Class. In England, it was quite different. Yes, they felt sorrow for the passengers, but the bulk of their sadness and despair was reserved for the crew. The loss of J. J. Astor, while important, paled in comparison to the loss of Thomas Andrews. In a time when duty meant everything, they did theirs, above and beyond the call. Still this was a tragedy shared. The great English humor magazine, "Punch," in its only mention of the *Titanic*, pictured *Britannia* and *Columbia* holding hands in their mutual grief but also in resolution.

Punch Magazine, April 24, 1912

> *"Tears for the dead, who shall not come again*
> *Homeward to any shore on any tide!*
> *Tears for the dead, but through that bitter rain*
> *Breaks, like an April sun, the smile of pride.*
> *What courage yielded place to others need,*
> *Patient of discipline's stern decree,*
> *Well may we guess who know that gallant breed,*
> *Schooled in the ancient chivalry of the sea."*

If we are speaking of publications, we must, [of course] include *The Social Register, The New York Times* of the upper class. Among its responsibilities was to keep track of what ship the wealthy and the élite were sailing on. There comes the dilemma. Madeleine Astor, Eleanor Widener, and the Duff-Gordons left Europe on the *Titanic*, but arrived in New York on the *Carpathia*. What to do? After much deliberation, they came up with a solution. Maggie Brown and seven hundred others crossed the Atlantic Ocean on the *"Titan-Carpath."* [Never in history did a hyphen have more meaning.]

As with any major disaster, an investigation would follow. This however, was the *Titanic*. Not one, but two governing bodies, three thousand miles apart, would each get a turn.

Chapter 13
Inquiring Minds

"It is not far from the truth to say that Congress in session is Congress on exhibition, whilst Congress in committee is Congress at work."

Woodrow Wilson

The *Carpathia* had reached New York, and the survivors of the *Titanic* were safe. The seven hundred men and women who walked down Cunard's gangway were something else, eyewitnesses. They had just witnessed the most notorious shipwreck of all time. The public [i.e. the newspapers] wanted answers. Well, if it's answers you want, who better to ask the questions than the United States Senate?

If ever there was a *Horatio Alger* story, it was the saga of William Alden Smith. At the age of twelve he moved with his family to Grand Rapids, Michigan. Shortly thereafter, he had to drop out of school; when his father became seriously ill with a lung disease. He helped support his family by selling popcorn, and working as a newsboy. At the age of twenty-one, he began studying law in the offices of Burch and Montgomery, paying his tuition in janitorial services. Smith was admitted to the Bar in 1882. Ten years later he was elected to Congress and served fourteen years in the House of Representatives, and then in 1906, he became a member of the United States Senate.

When news of the *Titanic* reached the Nation's Capitol, Smith went right to work. He pushed hard and got the Senate Commerce Committee to appoint a special investigation sub-committee with himself as chairman. Smith moved fast because he had no time to waist - the Navy had been intercepting messages from the *Carpathia:*

> "Most desirable Titanic crew aboard Carpathia should be returned home earliest possible moment. Suggest you hold Cedric, sailing her daylight Friday unless you see any reason contrary. Propose returning in her myself. Please send outfit of clothes, including shoes, for me to Cedric. Have nothing of my own. Please reply."
>
> *Yamsi*

Michel and Edmond Navratil, reunited with their mother

William Alden Smith, United States Senate

"Yamsi" was, of course, a clumsy anagram for Ismay. The man with no clothes had sent this message, along with two more, to Vice President Franklin in the New York office of White Star. Smith knew instantly what it meant. Ismay had plans for skipping town before being forced to answer questions. The Senator went right to the top, the White House. He asked President Taft, a former appeals court judge if there was any problem with placing a British citizen under a subpoena. The President replied, *"As long as it's on American soil, there is no problem at all."* He then requested a Treasury Department revenue cutter to intercept the *Carpathia,* thereby thwarting any chance of an Ismay escape. Taft instantly agreed.

With only minutes left, Smith made final arrangements and then raced to board the *"The Congressional Limited"* for the ride to New York. The train rolled into Penn Station at 9:07pm, and the committee immediately jumped into cabs for the trip to Pier 54. [Somehow managing to get through the traffic on West 14th Street.] Ten minutes after reaching the Cunard berth, they found the man they were looking for. White Star's Managing Director was served with a subpoena to appear before the committee the very next day.

At 9:00am, in the East Room of John Jacob Astor's Waldorf-Astoria, the New York portion of the Inquiry began. The old Michigan newsboy knew you never *"bury the lead."* The first witness gave his full name, Joseph Bruce Ismay. Ismay, dripping with sincerity made his opening remark. *"I would like to express my sincere grief at this deplorable catastrophe."* Smith's interrogation of the witness continued for some time and then gave the floor to Senator Newlands. Newlands, a Democrat from Nevada, got right to it: *"What was the full equipment of lifeboats for a ship of this size?"* The Senator had gone for the jugular and a bristling Ismay shot back:

J. Bruce Ismay, Senate Hearings, Waldorf-Astoria Hotel

> "I could not tell you that, sir...that is covered by the Board of Trade regulations. She may have exceeded the Board of Trade regulations for all I know."

[That was a bold face lie. Ismay had been well-schooled by both Carlisle and Andrews as to the boat situation.] He continued:

> "Anyhow she had sufficient boats to obtain her passenger certificate, and therefore she must have been fully boated, according to the requirements of the British Board of Trade, which I understand are accepted by this country."

[He had a point, but no one was willing to concede anything to him.]

In the next round of questioning Smith proceeded to grill the Managing Director about the ship's ability to float when damaged. Ismay replied:

> "The ship was specially constructed so that she would float with any two compartments full of water. I think I am right in saying that there are very few ships..."

Ismay paused, now was not the time to be quoting the Lines own propaganda, but he decided to say it anyway.

> "I believe there are very few ships today of which the same can be said. When we built the Titanic we had that especially in mind. If this ship had hit the iceberg stem on, in all human probability, she would have been here today."

On this point, Ismay was correct. There was no other ship that could have sustained the amount of damage as the *Titanic* and survived, and probably no other ship that could stay afloat for over two hours. Proof positive that the *"Irish Masters"* knew how to build a ship.

[What he neglected to include was that Ismay himself had nixed changes that would have made the *Olympic*-class liners safer still.]

The mood in the room changed when Smith called his next witness. A tall, distinguished man in a Captain's uniform walked in and sat down. For the record he gave his full name, Arthur Henry Rostron.

Smith guided the Captain through his testimony. Rostron described that having been informed that the impossible had happened to the unsinkable, he turned the *Carpathia* and headed north, ordering as much speed as possible. With those in the room silent and hanging on his every word, he went through his list of commands, and his all-out dash to the scene of the sinking. He was describing how, after rescuing the survivors, he made a search of the area but then Rostron paused in mid-sentence.

> "I want to go back again, a little bit. At 8:30am all the people were on board. I asked for the purser and told him that I wanted to hold a service..."

At that point, Rostron's voice broke down as his eyes filled with tears.

> "A short prayer of thankfulness for those rescued and a short burial service for those who were lost at sea."

By now Rostron was sobbing, as were many others in the room, including Senator Smith. The gallant Captain's sincerity had moved everyone. Tears were flowing again moments later when Smith finished his questions.

> "I think I may say for my associates that your conduct deserves the highest praise...and we are very grateful to you captain, for coming here."

The Senator and the Captain rose to shake hands, and there wasn't a dry eye in the house. A reporter [long on emotion, short on talent] had to resort to Rudyard Kipling to describe the moment.

> *"For there is neither East nor West,*
> *Border, nor breed, nor birth,*
> *When two strong men stand face to face,*
> *Though they come from the ends of the earth."*

The Captain of the *Carpathia* should be a tough act to follow, and so he was. The next witness was Second Officer Lightoller, who seemed more interested in protecting White Star than in giving useful information. His evasions and perfunctory answers gave minimal help to the committee. No doubt Lightoller felt insulted having to answer ship questions put to him by landlubbers. Meanwhile, Ismay was doing personal damage control with the American press.

> *"What do you think I am? Do you believe that I'm the sort that would have left that ship as long as there were any women and children on board? That's the thing that hurts, and it hurts all the more because it is so false and baseless. I have searched my mind with deepest care, I have thought long over each single incident that I could recall of that wreck. I'm sure that nothing wrong was done; that I did nothing that I should not have done. My conscience is clear and I have not been a lenient judge of my own acts."*

A parade of officers and crewmen followed Lightoller to the witness table. Fifth Officer Harold Lowe's attitude and answers were much the same as the Second Officer. No doubt, his contempt for and exasperation with the inquiry reached a peak when Smith inquired, "What is an iceberg made of?" Lowe calmly replied, *"Ice."*

Harold Bride being carried off *RMS Carpathia*

Officers Pittman & Lightoller, Senate hearings, New York

As *"W/O"* Harold Bride entered the room, one reporter commented, *"It was like an illustration of a gruesome story."* He arrived in a wheelchair, his head sunk deep in a pillow, his mouth slack and bandages covering his frostbitten foot. Though he looked sick, the operator's mind was sharp as a tack. [He was fit enough to help Cottam man *Carpathia's* wireless on the trip back to New York.] When he testified about the ice message from the *Californian* on the afternoon of the 14th the crowd gasped. This was the third warning that the liner had received. The press and guests, if not the Senators, were ready to dole out some blame.

Fourth Officer Boxhall was the first to mention the *"mystery ship."* Smith now doggedly pursued the name of the ship. He didn't have long to wait. The *Californian* had docked in Boston, and the crew was talking to the papers. A donkeyman [boiler maintenance] named Ernest Gill had signed an affidavit, testifying to what he saw. The next day it appeared in *The Boston American*. Below is the last portion of his statement.

> *"I am quite sure that the Californian was less than 20 miles from the Titanic which the officers report to have been our position. I could not have seen her if she had been more than 10 miles distant and I saw her very plainly. I have no ill will toward the Captain or any officer of the ship, and I am losing a profitable berth by making this statement. I am actuated by the desire that no captain who refuses or neglects to give aid to a vessel in distress should be able to hush up the men."*

Gill, the ship's engineers and her officers were all rushed to Washington and called before the Committee, where in turn, each corroborated the story.

U. S. Senate Inquiry, Washington, D.C.

P. A. S. Franklin and J. Bruce Ismay, U. S, Senate Inquiry

The papers had heard enough. They now had a "smoking gun." The man holding the pistol was Stanley Lord, Master of the *Californian.*

Lord testified that on the night of April 14th he saw a steamer coming up on her port side, well to the south. The Captain believed it was too small to be a liner, while the officer on watch, Third Officer Groves, was convinced it was. Lord told him to try and raise her with the Morse Lamp and went below. Groves studied the ship, saw her appear to turn her lights off and stop. An hour later, she was firing rockets. All of this was reported to the Captain.

The committee was stunned while Lord seemed oblivious to what was going on. He insisted that, given the *Titanic's* position, [Boxhall's] the *Californian* was 19 miles away, a distance that had been essentially refuted by the ship's previous witnesses.

The Captain's interrogation was remarkably brief. The next witness was the *Californian's* wireless operator Cyril Evans, who, having slept through the night had no pertinent information aside from telling the committee that the *Titanic* had told him to *"shut up"* as he was trying to send an ice-warning message.

More witnesses followed, including passengers [just two of them from Third Class] and "the great man himself," Guglielmo Marconi. At the conclusion of testimony Smith gaveled the hearings to a close. The Senior Senator from Michigan had some writing to do and a deadline to meet. Keep in mind that his committee had been charged with gathering evidence for new legislation. No one mentioned placing blame. [Try telling that to Smith, or Lord.]

The Senator's very long and somewhat flamboyant speech to the Senate is a book unto itself, but mercifully his recommendations for legislation, his conclusions, and observations can be stated easily.

[Some of the legislation was already in the "pipeline."] He had done a yeoman's job, pulling answers from reluctant witnesses in simple terms and not allowing them to obfuscate with technical jargon. On May 28th he made his address to Congress.

> *"No vessel shall be licensed to carry passengers from ports of the United States until all regulations and requirements of the laws of the United States have been fully complied with."*

Make no mistake, we have the authority, and will enforce our own laws and regulations." [And not just *"rubber-stamp"* the British].

> *"The committee recommends that sections 4481 and 4488, Revised Statutes, be so amended as to definitely require sufficient lifeboats to accommodate every passenger and every member of the crew."*

This is the one that the country, indeed the world was clamoring for, enough lifeboats for all.

> *"All members of the crew assigned to lifeboats should be drilled in lowering and rowing the boats, not less than twice each month and the fact of such drill or practice should be noted in the log."*

No more cursory examinations in port, or skipping drills on Sunday.

> *"The committee recommends that every ocean steamship carrying 100 or more passengers be required to carry 2 electric searchlights."*

This one sounds like winner until you realize that no ship had ever spotted an iceberg with a searchlight.

> *"The committee finds that this catastrophe makes glaringly apparent the necessity for regulation of radiotelegraphy. There must be an operator on duty at all times, day and night, to insure the immediate receipt of all distress, warning, or other important calls."*

Simply put, the world could no longer come off its hinges while Cyril Evans slept in his bunk.

> *"The committee recommends that the firing of rockets or candles on the high seas for any other purpose than as a signal of distress be made a misdemeanor."*

There was no mention of any punishment for ignoring a distress signal. Captain Lord had skated on that one.

> *"All steel ocean and coastwise seagoing ships carrying 100 or more passengers should have a watertight skin inboard of the outside plating, extending not less than 10 percent of the load draft above the full-load waterline, either in the form of an inner bottom or of longitudinal watertight bulkheads, and this construction should extend from the forward collision bulkhead over not less than two-thirds of the length of the ship."*

The Committee may not have realized it, but they had re-affirmed the brilliance and vision of a man whose body was at the bottom of the ocean, a thousand miles away. It had taken a still unbelievable disaster to vindicate Thomas Andrews.

> *"All steel ocean and coastwise seagoing ships carrying 100 or more passengers should have bulkheads so spaced that*

any two adjacent compartments of the ship may be flooded without destroying the floatability or stability of the ship."

The irony? *RMS Titanic* met their recommendations and still sank. [*RMS Mauretania*, built to Admiralty standards, would have as well.]

The Senate had finished its work without taking anyone out to the wood shed. The press, however, would draw down on two targets of opportunity. Stanley Lord would take a pounding, but since Edward Smith was dead, his bombs would be dropped on J. Bruce Ismay.

There was a difference of opinion as to the quality of Smith's Inquiry. The American press lauded the Committee for its thoroughness and recommendations. The British opinion was summed in the words of Lightoller, who, in his autobiography described the American Inquiry as *"a farce"* due to the ignorance of maritime matters implicit in some questions. The English press labeled the Michigan Senator *"Watertight Smith"* for asking whether watertight compartments, actually meant to keep the ship afloat, were meant to shelter people.

Smith, however, was not the simple-minded bumpkin that the boys of Fleet Street made him out to be. He had toured *RMS Olympic* and knew precisely what the watertight bulkheads did, but understood that the general public might not. Questions like that were designed to help citizens and were asked with the highest of motives. Regardless, the English reporters had hit him in his soft spot - his altruism. The Senior Senator from Michigan reloaded and fired the last cartridge in his belt. Not included in the report was a statement made to the press about the times they lived in.

"What this nation needs is a severe lesson that will strengthen the pillars of its faith. We are running mad with the lures of wealth, of power, and of business. We are

setting society into castes, with the forces of wealth and power on one side and destitution and poverty on the other. It takes a terrible warning to bring us back to our moorings, and to our senses."

Well, as they say at Wimbledon, the ball is in your court. Next would come the British Investigation into the wreck of the *Titanic*. Ah, but this one came with a twist. The very same people who set the rules, The Board of Trade would be the ones in charge of the investigation. [Akin to a judge killing a man and then presiding over his own trial.] The Inquiry opened on May 2, 1912, in Westminster, with Lord Mersey [1st Viscount Mersey], serving as the Wreck Commissioner.

In all, 101 witnesses would give testimony over the course of 36 days. It would have helped if there had been three more. First of all, Captain Edward Smith could explain why his ship was doing above 22 knots in spite of numerous ice warnings, while First Officer William Murdoch could explain why he chose to maneuver *Titanic* around the ice, and Thomas Andrews could [perhaps] answer the question all of England was asking. [To paraphrase Admiral Beatty] "What went so bloody wrong with our ship?" The Board set to work:

> "On or about the 14th April, 1912, the British Steamship 'Titanic,' of Liverpool, struck ice in or near latitude 41° 46' N, longitude 50° 14' W, Atlantic Ocean, and on the following day foundered and loss of life thereby ensued or occurred."

On Day 3, everyone's favorite lifeboat commander was examined. Quartermaster Robert Hichens gave a thoroughly lucid and equally self-serving account of his time spent in Lifeboat No. 6. Over the next several days the court took testimony from surviving officers and crewmembers, including a now sober baker Joughin.

Lord Mersey, Wreck Master, Board of Trade Inquiry

QM Robert Hichens testifying at Board of Trade Inquiry

On Day 7, the court examined witnesses from the *Californian*. Before he even took the stand, Captain Lord was in the crosshairs:

> "The *Californian* is said by this donkeyman [Ernest Gill] to have seen the distress rockets fired from a vessel which, according to this man, was the *Titanic* and to have taken no notice of those distress rockets. There is no doubt, as I understand the evidence, that rockets were seen on this night and that the *Californian* was not at a very great distance from the *Titanic*."

Captain Lord testified that, based on positions, he judged the *Titanic* to be nineteen miles away. The ship he saw from the bridge was about six miles away and definitely not the *Titanic*. He answered his questions in a direct and forthright manner. It was clear that he believed what he said, and just as clear, the court members did not.

The next day, Third Officer Groves repeated the story that was in the Boston newspapers about how he saw the lights of a large steamer 10 to 12 miles away with deck lights on. Gill's testimony was consistent with what he told the reporters, but it was Groves holding the dagger.

> *8441. Speaking as an experienced seaman and knowing what you do know now, do you think that steamer that you know was throwing up rockets, and that you say was a passenger steamer, was the "Titanic"?*
>
> Groves: Do I think it?
>
> *8442. Yes?*

> *Groves: From what I have heard subsequently?*
>
> *8443. Yes?*
>
> *Groves: Most decidedly I do, but I do not put myself as being an experienced man.*
>
> *8444. But that is your opinion as far as your experience goes?*
>
> *Groves: Yes it Is, my Lord.*

Now the Inquiry turned away from the *Californian* and to the subject of lifeboats. Sir Cosmo Duff-Gordon testified that his wife and her secretary refused to get into Lifeboats No. 5 and No. 3 without him, and then all three of them climbed into No. 1. The Board then probed their behavior in the cutter. Sir Cosmo and his wife had been plagued by rumors that they bribed the crew in the lifeboat not to pick up survivors and to row away from the wreck. Since any male survivor was now suspect, he fought hard to save his reputation. Apparently the Board liked what it heard and cleared them of any wrongdoing. Duff-Gordon, his reputation somewhat restored left the hearing.

Senator Smith chose to put Ismay on the stand the very first day. The Board was in no hurry and waited until Day 16. White Star's Managing Director was co-operative at first, but, as the day wore on, he reverted to form. Ismay's evasions, which did not get him far at the Waldorf, got him nowhere at all in Westminster. This exchange followed his claim not to know what latitude and longitude meant.

> *18428. And you knew also that you would be approaching ice that night?*

J. Bruce Ismay, Board of Trade Inquiry

Titanic crewmen, Board of Trade Inquiry

Ismay: I expected so, yes.

18429. *And that you therefore would be crossing the particular region, which was indicated in that Marconigram that night?*

Ismay: I could not tell that.

18430. *About that region?*

Ismay: Yes, I presume so.

18431. *And therefore that it behoved those responsible for the navigation of the ship to be very careful?*

Ismay: Naturally.

18432. *And more particularly if you were approaching ice in the night it would be desirable, would it not, to slow down?*

Ismay: I am not a navigator.

18433. *(The Commissioner) Answer the question.*

Ismay: I say no. I am not a navigator.

The Attorney-General: You are not quite frank with us, Mr. Ismay.

It continued like this for two full days. Ismay was at times co-operative, other times not, but always he responded in the best interest of Ismay. It was not lost on the court that potentially helpful witnesses [Smith, Murdoch and Andrews] were not available, while the Inquiry had to make do with a less than helpful, and at times a down right hostile witness - Ismay. Three days later the original designer, Alexander Carlisle was summoned to appear. He found a way to make his former employer [Harland & Wolff] look bad. The *"Sanderson"* noted below is IMM Vice President Harold Sanderson. The subject is the plans for the *Olympic* and the *Titanic*.

> *21289. Did Mr. Sanderson discuss those plans?*
>
> *Carlisle: Mr. Sanderson, I think, never spoke.*
>
> *21290. Did he sit for four hours without speaking?*
>
> *Carlisle: No, but that was over the whole of the decorations; we took the entire decorations of that ship.*
>
> *21291. Never mind about the decorations, we are talking about the lifeboats?*
>
> *Carlisle: The lifeboat part I suppose took five or ten minutes.*
>
> *21292. Then, am I to understand that these plans which you are now producing were discussed, at this four hours interview for five or ten minutes?*
>
> *Carlisle: That is so.*

As noted, the Board took thirty-six days of testimony before releasing their report. They had heard from passengers and crew as well as an array of expert witnesses. M'Lord Mersey did not mince words.

> "The Court, having carefully inquired into the circumstances of the above mentioned shipping casualty, finds, for the reasons appearing in the annex hereto, that the loss of the said ship was due to collision with an iceberg, brought about by the excessive speed at which the ship was being navigated."

The Inquiry laid the blame at the feet of a man who wasn't alive to defend himself, Edward Smith. The Report went on to recommend changes in the rules, [the Board's own rules] before coming to the big one: *"The Circumstances in Connection with the Californian."*

> "These circumstances convince me that the ship seen by the Californian was the Titanic, and if so, according to Captain Lord, the two vessels were about five miles apart at the time of the disaster. The evidence from the Titanic corroborates this estimate, but I am advised that the distance was probably greater, though not more than eight to ten miles. The ice by which the Californian was surrounded was loose ice extending for a distance of not more than two or three miles in the direction of the Titanic. The night was clear and the sea was smooth. When she first saw the rockets the Californian could have pushed through the ice to the open water without any serious risk and so have come to the assistance of the Titanic. Had she done so she might have saved many if not all of the lives that were lost."

It had been sixty years since Gilbert and Sullivan's *"The Mikado"* had opened in the West End of London. In it the Lord High Executioner sings how, for people he doesn't approve of, *"He's got a little list."* Assuming that Lord Mersey and the Board of Trade had such a list, the name at the top would be Stanley Lord.

In the Inquiry's final report there was no admonishment of the Board itself, whose pathetically inadequate rules were, in large part, responsible for the appalling loss of life. To no one's surprise, the British press praised the Board of Trade for a *"serious"* Inquiry. Charles Lightoller, who answered no fewer than 1600 questions, portrayed the role of the "company man" to the hilt.

> *"It was very necessary to keep one's hand on the whitewash brush...I had no desire that blame should be attributed either to the Board of Trade or the White Star Line."*

The "blame assignment phase" continued on both sides of the Atlantic, with the British papers castigating the United States for its avarice and Americans going after the English for their arrogance.

In the weeks following the wreck, the *Olympic* assisted both the American and British Investigations into the disaster. Deputations from both Inquiries inspected her lifeboats, watertight doors and bulkheads and other equipment, which were identical to those on *Titanic*. Sea tests were performed for the Board of Trade in May 1912, to establish how quickly the ship could turn two points at various speeds, and to discover how long it would have taken the *Titanic* to turn when it sighted the iceberg. At 22 knots it would take 37 seconds. It seemed the *"Gilded Age Masters of the Universe"* had missed another one.

Of all the adjectives lavished on the *Titanic,* the one that never entered the conversation was "agile." The ship was, in part because of design and in part because of her displacement [50,000 tons], a poor handler. For all her style and elegance, she was rather clumsy. [Picture Ginger Rogers, now picture her with two left feet.] The other liners of her era [if only because they were smaller]

"Nearer My God to Thee," Toledo News

could out maneuver her, turn more easily and in a tighter circle. In time, Marine Architects would make major discoveries about a vessel's movement through the water. They learned the sharpest turn is made when the engines are full ahead and the rudder is hard over. [A ship turns when force is applied to the rudder.]

In the language of Baseball, it was three strikes and you're out.

Strike One: By virtue of a center propeller, the rudder was not big enough for the size of the ship.

Strike Two: The wing propellers were reversed. The slight decrease in speed would be overwhelmed by the loss of turning ability.

Strike Three: The center propeller, since it could not be reversed, was stopped. The prop closest to the rudder would exert no force.

Lest you think that the blame for the wreck should be laid at the feet of the builder – consider this. In spite of horrific damage that would have sunk another liner in far less time, she lived for nearly three hours. At the dawn, sixteen yard-built lifeboats, some loaded above capacity, would bring the survivors safely to the *Carpathia's* ladder. The *Irish Masters* of Harland & Wolfe had kept faith with the *Titanic*.

In 1914, in response to the sinking of the *RMS Titanic*, the International Ice Patrol was formed. The Patrol's mission was to alert any craft traveling the great circle lanes between Europe and the major ports of the United States and Canada of the presence of any icebergs. Now, the hunters would become the hunted.

Every year since 1914, on the 15th of April, the Coast Guard or International Ice Patrol ship on duty off the *Grand Banks* will steer a course for 41' 46" N, 50' 14" W, and there, lay a wreath on the water in remembrance of the fifteen hundred souls who perished that night.

Captain Smith Memorial, Staffordshire, England

Chapter 14
Aftermath

*"They that go down to the sea in ships,
that do business in great waters;
These see the works of the Lord
and his wonders in the deep"*

Psalms 107:23

"There was peace and the world had an even tenor to its way. Nothing was revealed in the morning the trend of which was not known the night before. It seems to me that the disaster about to occur was the event that not only made the world rub it's eyes and awake but woke it with a start keeping it moving at a rapidly accelerating pace ever since with less and less peace, satisfaction and happiness. To my mind the world of today awoke April 15th, 1912."

Jack B. Thayer

Master Thayer had nailed it. [Quite astute for a seventeen-year-old.] Various things started "hitting the fan" almost immediately. The *Olympic* arrived at Southampton in the early hours of Sunday, April 21st to find the port in a state of mourning. In spite of the all-pervading gloom, activity in the dockyard was more intense than usual. Berthed at the White Star Dock [Pier 44], she had to be prepared for her next scheduled departure, due three days later. In the light of recent events, considerable modifications were to be made to the life saving apparatus on the Boat Deck.

The yard embarked forty Berthon collapsible lifeboats, some of which were tested under the watchful eye of Captain Maurice Clarke, the man who had cleared *Titanic* for her sailing less than two weeks earlier. Just prior to departure, Captain Haddock was handed a message. It seemed that his stokehold crew had found a reason for deserting the ship. Further tests revealed that five of the Berthons were totally rotten and two others leaked. It took a full day, another mutiny, and numerous negotiations before enough of the crew was back on board to finally sail, but the sailors had made their point. The White Star and all the other shipping lines, which had provided the bare minimum in life saving equipment, would do more than pay lip service to the new regulations.

The fourth estate was now running amuck. Every reporter with a press pass [and in the case of *The New York Times*, many without] had met the *Carpathia* at the dock. All had come back with a tale to tell. Survival stories filled the newspapers, in the process pushing the Presidential Campaign of 1912 and everything else to the back pages. On April 20, the opening of the new home of the Boston Red Sox, Fenway Park, could do no better than page six of *The Boston Globe*.

The feud between J. Bruce Ismay and William Randolph Hearst went back a long way. Ismay had met Hearst years before, when he was White Star's agent in New York. The two men disliked each other intensely, and Ismay's refusal to cooperate with the press infuriated the newspaperman. Hearst had stockpiled his resentment and waited for his chance. Ismay was everything he hated, an arrogant snob who had been born into money. The fact that Hearst was just as arrogant, as snobbish, and heir to a fortune was beside the point. *"Never pick a fight with anyone who buys his ink by the barrel"* [Thank you, Mark Twain]. The waiting period was over. Hearst now, *"let loose the hounds"* with himself *"the leader of the pack."*

> *"Who would not rather die a hero than live a coward? These men have died as men should die. They performed their duty to their fellow men, their obligation to their God. So may God reward them and men remember them. And may the memory of them remain forever a noble record of past heroism for humanity, a splendid inspiration to future deeds of duty and devotion."*
>
> *William Randolph Hearst*

William Randolph Hearst, publisher

Ben Hecht, reporter, *Chicago Journal*

The *"Chief"* [Hearst] may not have known it, but he had touched a nerve. The British Inquiry had exonerated Ismay for leaving in a lifeboat. The Board of Trade decided that if he had remained on deck, it would have done nothing more than add one name to the death toll. As many a celebrity would discover over the next century, it was the Court of Public Opinion that would be the ultimate arbiter.

The public needed a scapegoat, and Hearst handed them Ismay on a silver platter. His readers, hungry for information about the disaster and eager to find someone to blame, were only too happy to accept his gift. Countless wire stories asserted Ismay's guilt of manipulating the *Titanic's* master into driving his ship faster than he wanted; of cowardice in taking the place of a passenger in a lifeboat; and of resigning from the company after the disaster rather than face the public. None of this was true, but Hearst, like many a newsman before and since, *"Never let the facts get in the way of a good story."* [Some called him the *"Coward of the Titanic"* or *"J. Brute Ismay,"* while others wanted White Star's pennant changed to a yellow liver.] A cub reporter named Ben Hecht deftly summarized the Hearst position in the April 17th edition of the *The Chicago Journal*.

> **"To hold your place**
> **in the ghastly face of**
> **death on the sea at night**
> **is a seaman's job,**
> **but to flee with the mob,**
> **is an owner's noble right."**

King Edward VII died in May of 1910. The era that was named for him was already starting to wane. The end would come four years later with Maxim machine guns slaughtering English and French soldiers in the trenches by the Marne. This period was both one of great social change and of solidifying the power and luxury of the ruling élite. The gap between rich and poor had become a chasm. In 1912 England, over a million people worked as servants to the upper classes. This dichotomy was indentified by two [French] phrases, *"La Belle Epoque"* [the beautiful epoch] and *"Fin de siècle"* [end of a century]. The German summation came in one word, *"zeitgeist"* [spirit of the times]. In few other eras in history were the times this "spirited." With the possible exception of the Roman Empire, no other period has witnessed such decadence and pessimism, optimism and hope. Society was no longer a small, exclusive circle confined to those of aristocratic birth. If you had money and followed the rules you could work your way in. In the Edwardian era there had been "guidelines" for everything: how to dress [leave un-buttoned the bottom button of your waist coat]; dinner [black tie instead of white]; and marriage [better a mistress than a divorce]. You could buy anything, even a title. *"Dollar Princesses"* [daughters of American millionaires] were now marrying impoverished Dukes, thereby exchanging *"Yankee greenbacks"* for a place among the nobility.

It would be the last hurrah of the aristocracy. The seeds of Nationalism had taken hold and would, in time, spell the doom of the British Empire and with it the *Pax Britannica*. The morning of a new day was coming, and while the *Titanic* was no longer on the front page, she would still be a part of it. To paraphrase a 1960's balladeer, *"the times they were a-changin."*

Morning, however, would not arrive until the world was finished mourning. The cable repair ship *Mackay-Bennett* quietly slipped out of Halifax on the evening of April 15. In her holds the Canadian ship was carrying a cargo of coffins and dry ice.

King Edward VII in his Coronation Robes

Canadian Cable Ship, *CS Mackay-Bennett*

***Titanic* Memorial Service, St. Paul's Cathedral**

Wallace Hartley Memorial, Colne, England

She was contracted by White Star to carry out the sorry task of recovering the bodies left floating in the Atlantic after the disaster. She sailed with no one to see her off, now, two weeks later she returned with all of Halifax standing quietly on the shore. On board were 190 bodies [a further 116 buried at sea] including #15 Michel Navratil, #96 Isidor Straus, #124 John Jacob Astor IV, and #137 Chief Purser Hugh Walter McElroy. Closure could now commence.

Heroes were not hard to find. The papers carried stories of the brave and the selfless. The sad finish of the *Titanic's* orchestra broke hearts around the world. How those eight musicians kept playing until the end. The *Mackay-Bennett* recovered the body of Wallace Hartley ten days after the sinking. Hartley was laid to rest in his hometown of Colne on May 18, 1912. One thousand people attended his funeral while forty thousand more lined the route of his procession. A ten-foot headstone, containing a carved violin at its base, was placed on his grave. A memorial to him was erected in 1915 outside the Methodist Church in Colne where Hartley began his musical career.

The crowd at Archibald Butt's funeral was smaller but no less distinguished. William Howard Taft, the President of the United States, delivered the eulogy.

> *If Archie could have selected a time to die he would have chosen the one God gave him. His life was spent in self sacrifice, serving others. His forgetfulness of self had become a part of his nature. Everybody who knew him called him Archie. I couldn't prepare anything in advance to say here. I tried, but couldn't. He was too near me. He was loyal to my predecessor, as his military aide, and to me he had become as a son or a brother.*

Wallace Hartley's funeral, Colne, England

John Jacob Astor's funeral, Rhinebeck, New York

John Jacob Astor, in life just another millionaire, was in death, someone to be revered. His caring for his pregnant wife and then resigning himself to his fate tugged at the heartstrings of readers. Plans for his funeral rated a front-page story in *The New York Times*

Services at Rhinebeck Today
Burial in Trinity Cemetery Here

RHINEBECK May 3 *The funeral service of Col. John Jacob Astor will be held tomorrow at noon in the Episcopal Church of the Messiah, of which he was Warden for sixteen years. A special Train from New York bearing friends and relatives is due to reach Rhinecliff Station at 11:20 o'clock. Automobiles will be in waiting to convey the passengers to the church.*

A private train was not required to attend the funeral of Isidor Straus, the subway would do. [What is now the "J" train to Cypress Hills.] If heroism rang true in the Edwardian era, so did undying love. The saga of Isidor and Ira refusing to be separated moved a generation. The remains of Mr. Straus were first buried in the Straus-Kohns Mausoleum at Beth-El Cemetery. His body was moved to the Straus Mausoleum in the Woodlawn Cemetery in the Bronx in 1928. Isidor and Ida are remembered on a cenotaph with words from *"Canticles."*

> "Many waters cannot quench love - neither can the floods drown it."

There is also a plaque on the main floor of *Macy's* Department Store in Manhattan and their sons would bequeath *Straus Hall,* one of the freshman residences in Harvard Yard, in honor of the their parents.

Memorials sprouted on both sides of the Atlantic. The *Titanic's* engineers, all of whom had perished, were honored in two cities.

> Owing to the death of
> Mr. and Mrs. Isidor Straus
> this store is
> closed to-day, Saturday.
>
> *R. H. Macy & Co.*
>
> HERALD SQUARE,
> Broadway, 34th to 35th St.,
> NEW YORK.

"Macy's" notice, *The New York Times*

Straus Memorial, Broadway and W. 106th Street, New York

Titanic **Engineer's Memorial, Liverpool, England**

A stone obelisk rose above Liverpool, while a bronze and granite monument graced Southampton. The inscription reads:

**TO THE MEMORY OF THE ENGINEER OFFICERS
OF THE R.M.S "TITANIC" WHO SHOWED
THEIR HIGH CONCEPTION OF DUTY AND THEIR
HEROISM BY REMAINING AT THEIR POSTS
15TH APRIL 1912.**

Thomas Andrews was remembered by his fellow countrymen with the construction of a Memorial Hall in Comber, County Down. The money to fund the project came from donations by his own men, the shipwrights of Harland & Wolff. The first sod was cut by his daughter, Elizabeth Law Barbour Andrews [ELBA], and opened a year later by his widow Helen. The building still stands and is now used as a primary school. It is safe to say that every student knows about the legend of Thomas Andrews.

In later years, a shadow would fall on him [undeservingly] but not in his hometown of Dalbeattle, Scotland. William McMaster Murdoch was honored with a large plaque on the town hall; it *"commemorates the heroism"* of the *Titanic's* First Officer. The town also established a memorial fund in his name. In 1997, an American film director, with only the highest motives, donated £5,000 to the fund. Dalbeattle had already received what they wanted most - an apology.

Hartley's musicians would be memorialized in both Liverpool and Southampton, just as the ship's engineers had been. A statue was erected to Captain Smith in his hometown of Litchfield, England, and another for Archibald Butt in Washington. There is a steward's monument in Southampton and monuments to all in New York, Belfast, Liverpool, and Queenstown.

Women's Monument to the Titanic, Washington, D.C.

As you might expect, the New York memorial was unique. No plaques or statues for the people on the Hudson, they built a lighthouse. Erected by public subscription in 1913, it stood above the East River on the roof of the old Seamen's Church Institute of New York. The 60-foot-tall structure, built in large part due to the instigation of Margaret Brown, was meant to be a constant reminder of the lives lost on board the *Titanic*.

There is simple granite statue in a quiet corner of Washington, D.C. It depicts a man with his arms outstretched, erected by the Women of America. Below it are the words:

> TO THE BRAVE MEN
> WHO PERISHED
> IN THE WRECK
> OF THE *TITANIC*
> APRIL 15 1912
> THEY GAVE THEIR
> LIVES THAT WOMEN
> AND CHILDREN
> MIGHT BE SAVED

The world would change after the loss of the *Titanic,* but it would be a slow torturous process. Of all the coverage that abounded after the wreck, one of the most famous [and telling] articles was, from of all places, *The Wall Street Journal*. *The Journal* had bitten hard on the bogus story that all the passengers had been saved. The situation demanded homage to the wonderfulness of man still prevailing no matter what was put in his path. When the edition of April 16, 1912 went to press, the *"op-ed"* section included the following:

"The gravity of the damage to the Titanic is apparent, but the important point is that she did not sink. Her watertight bulkheads were really watertight. Man is the weakest and most formidable creature on the earth. His physical means of protection and offense are trifling. But his brain has within it the spirit of the divine and he overcomes natural obstacles by thought, which is incomparably the greatest force in the universe."

[Hubris dies hard.] How many saw that opinion is open to debate. *"The Journal,"* like all other papers was now irrelevant. The *"story of the century"* belonged to Carr Van Anda and *The New York Times*.

Regardless, no one was going to recklessly drive a ship at full speed through an ice field. The concept of an "unsinkable" ship was gone forever. Along with it came the notion that the whole thing was inevitable, that Robertson's now popular book was right. Lightoller's statement: *"the circumstances might not occur for a century,"* seemed lost against the fact that it could...and did.

The sociology of the world was changing in a profound way. Privilege [pronounced *"money"*] was not what it used to be, not when you can pay 4,000 quid for a Parlor Suite and be denied access to a lifeboat. Lower class passengers, who had been told for generations that there was no way up for them, had seen it play out in a very real, deadly way. That would change as well. No one would ever again lock doors below decks to deny less affluent passengers a chance to be saved.

And, while the *Titanic* would help to advance many causes, it would also set one back. The Women's suffrage movement took a direct hit. The cry of *"votes for women"* lost some of its zing when placed along side, *"women and children first."* The sinking had revealed what some women had chosen to forget. Equal rights brings with it equal risk and equal responsibility.

All manner of theorists and poets would take a shot at what had happened and what it meant. I like Thomas Hardy's version. Below is his *"Convergence of the Twain"* (*"Lines on the loss of the Titanic"*), an English poet's explanation of how it all went so terribly wrong:

> *In a solitude of the sea*
> *Deep from human vanity,*
> *And the Pride of Life that planned her, stilly couches she.*
>
> *Steel chambers, late the pyres*
> *Of her salamandrine fires,*
> *Cold currents thrid, and turn to rhythmic tidal lyres.*
>
> *Over the mirrors meant*
> *To glass the opulent*
> *The sea-worm crawls — grotesque, slimed, dumb, indifferent.*
>
> *Jewels in joy designed*
> *To ravish the sensuous mind*
> *Lie lightless, all their sparkles bleared and black and blind.*
>
> *Dim moon-eyed fishes near*
> *Gaze at the gilded gear*
> *And query: "What does this vain gloriousness down here?"*
>
> *Well: while was fashioning*
> *This creature of cleaving wing,*
> *The Immanent Will that stirs and urges everything*
>
> *Prepared a sinister mate*
> *For her — so gaily great —*
> *A Shape of Ice, for the time far and dissociate.*
>
> *And as the smart ship grew*
> *In stature, grace, and hue,*
> *In shadowy silent distance grew the Iceberg too.*

Alien they seemed to be;
No mortal eye could see
The intimate welding of their later history,

Or sign that they were bent
By paths coincident
On being anon twin halves of one august event,

Till the Spinner of the Years
Said "Now!" And each one hears,
And consummation comes, and jars two hemispheres.

The *Titanic* was, if nothing else, the precursor of the great social bulldozer of the early twentieth century, World War I. Suddenly the loss of fifteen hundred lives seemed less significant when balanced against a generation of Frenchmen gunned downed around Verdun. It would take a World War to push *Titanic* off the front page. The World that emerged in 1918 was very different from the one that had gone to war four years before. And, while interest in the events of April 15, 1912, waned, she still managed to become part of the American conscience and folklore.

Exhibit 1: The verse and one chorus of arguably the most popular camp song of all time.

"Oh, they built the ship Titanic,
to sail the ocean blue.
For they thought it was a ship
that water would never go through.
It was on its maiden trip,
that an iceberg hit the ship.
It was sad when the great ship went down.

It was sad, so sad.
It was sad, so sad.

> *It was sad when the great ship went down*
> *(to the bottom...)*
> *Uncles and aunts,*
> *little children lost their pants.*
> *It was sad when the great ship went down."*

[Since I wanted no part of camping, the version I learned said *"husbands and wives, little children lost their lives."* Either works.]

In the three generations since the abolition of slavery, the struggle for equality had not gone well for black men and women. About the best job a black man could get was as a Pullman porter on a long distance train. The dominant sport of the time, baseball, was closed to men of color. Commissioner Landis and many of the owners, and players for that matter, were racist through and through. You could not serve in the Army; you could only serve as a cook or a stevedore in the Navy. Carefully mix in one Great Depression and you have anger and resentment. So, it was with no small amount of glee when this little ditty came through the projects.

Exhibit 2: A song where a lot of white people die and a black man lives. History would remember it as the *"Titanic Toast,"* and it concerns a black stoker named "Shine." And like any good folk song, certain "liberties" are taken with the facts. [The eighth of May?]

> *"You know the eighth of May*
> *Was a hell of day!*
> *When the Titanic struck the big iceberg.*
>
> *Up jumped Shine from two decks below,*
> *He said, 'Captain, Captain, the waters*
> *All over the boiler room floor.*

> *Captain look at him, told him, say,*
> *Go back down there, Shine and pack your sacks*
> *Say, I got ninety-nine pumps on here to keep the water back."*

Shine does what he is told, but on the way down he sees the water rising, so he jumps overboard and starts to swim.

> *"So the Captain told Shine, say*
> *Come back here, Shine, and save poor me,*
> *He say, I'll make you rich as any Shine can be."*

Ignoring offers of cash from white men and offers of sexual favors from white women, Shine swims all the way back to New York.

> *"And guess where he was*
> *When the news reached the seaport town*
> *That the Titanic had sunk?*
> *He was over in Edgewood, dead drunk!"*

The *Titanic* is everywhere: Movies, plays, poems, books, doggerel, television, musicals and magazines. Maybe the *Shipbuilder* was right; maybe she was "unsinkable."

Chapter 15
The "L" Word

*"People are getting smarter nowadays;
they are letting lawyers,
instead of their conscience,
be their guide."*

Will Rogers

The American Bar Association's annual meeting for the year 1998 was held in Toronto, Canada. Surely the highlight of the event was a one-day mock trial. In this trial, a mythical survivor named Rhoda Abbott [more about her later] had filed a lawsuit in regards to the loss of her husband aboard *RMS Titanic*. Her children had survived. Mrs. Abbott's complaint stated the following facts:

> *The ship was unsafely built and operated*
> *That the rivets were brittle*
> *The ship itself was not easily maneuverable*
> *There were not enough lifeboats*
> *That the lookouts didn't have binoculars*

The suit was not filed against the operators - The White Star Line, nor the owners - International Mercantile Marine, but rather the shipbuilders - Harland & Wolff. Lead counsel for the Plaintiff was Johnny Cochran. [Yes, that Johnny Cochran.]

The Plaintiff's lawyer told the court that, *"If you design or build a boat, it must be able to stay afloat."* [Cochran's rhyme] The Defendant's lawyer, Stephan G. Morrison, countered by saying his client had built *"a state of the art masterpiece."* Like any good defense lawyer would, he tried to deflect blame from his client with an *"alternative theory of the crime"* by claiming that the real culprit was *"the ship's owner and operator, the White Star Line."*

The trial was conducted using contemporary rules with both sides calling witnesses to bolster their case. Then, each made a closing statement, with Cochran asking for anywhere from $120,000,000 to $225,000,000 in damages. The jury [the audience] deliberated and came back with a verdict. They found for the Plaintiff.

Holding Harland & Wolff liable for $1,500,000 in damages.

"If the ship ain't stout, you must pay out." [My rhyme]

By the way, there was nothing "mythical" about Mrs. Abbott. Rhoda Mary Abbott, you may recall, was a divorcee traveling in Third Class with her two sons. She made it into the "A" boat, but her sons perished in the wreck. Did the lawyers know she was not "mythical," or did the newspapers and television just get the story wrong? In any event, Rhoda Abbott finally got her day in court, and armed with 1998 law, a 1998 jury [and Cochran], the Plaintiff finally prevailed. [Maybe, just maybe, Shakespeare was wrong?]

We live in a litigious society, but we hardly invented the concept. In 1912, there was no shortage of attorneys and an abundance of litigation. [The "L" word] What if you were an immigrant family and your breadwinner had just been swept away, what happens next? Benefits could help out in the short run, but what about the rest of your life? In the lexicon of the law, these people had all been *damaged* and were entitled to *compensation.*

Let's suppose you were a woman with young children, now a widow as a result of the disaster, and you want, as Shakespeare would say, your *"pound of flesh."* How would you go about it? What were your options? [Keep in mind that the following is all supposition.]

You could be from First, Second, or Third Class, but, for the sake of the newspapers and the publicity, we'll go with First. Our Plaintiff is Eleanor Widener. Mrs. Widener watched from a lifeboat as her husband and son drowned. Now, did the widow of one of the richest men in Pennsylvania need the money? Hardly. Nevertheless, someone was going to pay dearly for the deaths of George and Harry. So, let the games biegin.

Survivor Benefit, *RMS Titanic*, April, 29, 1912

Thus, armed with a cadre of sharp "Philly" lawyers, Mrs. Widener filed and served her complaint. The process started with a determination of *"locus standi"* [standing]. For those of us who did not attend law school, *standing* is defined as:

> *"The ability of a party to demonstrate to the court sufficient connection to and harm from the law or action challenged to support that party's participation in the case."*

Since Eleanor was the wife of George and the mother of Harry, since this was an American court, and since IMM was American-owned, it would have been pointless for the Defense to challenge her on this basis. Eleanor Widener had standing by the ton.

The complaint would list causes of action. In this case, the most likely would be "negligence." The lawyers could pile on other things, such as "loss of consortium." If the year was 2012, rather than 1912, they could probably make a case for sexual discrimination, since the order to fill the lifeboats was *"women and children first."* The complaint would also list facts in support of the causes of action, and would be followed by a claim for damages. *"Let the games continue!"*

Our mythical complaint would be one of numerous lawsuits filed to recover for loss of life and personal injury against the ship owner, White Star Line, in Federal and State courts in New York. It must have come as a shock to the Plaintiffs when a petition was filed by the *"foreign vessel owner"* in the U.S. District Court for the Southern District of New York to limit its liability under United States law.

[Supposition ends right here.] In 1851, Congress, *"to encourage investments in ships and promote a healthy U.S. Merchant Marine,"* passed what they thought would be a "helpful" law.

Under the *"Limitation of Liability Act,"* an owner could limit their liability if they could prove that the loss was occasioned without its *"privity"* [translation *"knowledge"*] such as an error in navigation or, in this case, the result of an Act of God [translation *"iceberg"*].

Since the *Titanic* sank after striking the iceberg, the Defense claimed that the lifeboats were the owner's only remaining assets. *"Really?"* This assumes you don't count *Olympic,* the scantlings of *Gigantic* or any of the other thirty liners and tenders owned by White Star. Thus, under the U.S. Limitation Act, the owners sought to limit any recovery to the meager value of thirteen lifeboats and freight money earned. The grand total came to something less than $92,000.

The Plaintiffs immediately countered with a motion to dismiss the owner's limitation petition, arguing, *"that no wording in the Limitation Act made it applicable to a foreign corporation."* The argument continued, and, after two years of appeals, the case wound up in the lap of the Supreme Court. The Court ruled for the Defendants, having determined:

> *"That in the case of a disaster upon the high seas with claimants of many different nationalities pursuing remedies here, the foreign owner may indeed pursue a limitation of liability proceeding under U.S. admiralty law."*

[So it would seem, families were not the only ones with *"standing."*] The intent of the law, of course, was to protect American Companies and certainly not British. Once again the Supreme Court had overreached. With this decision, the Defendants averted another disaster, not so the passengers. Had the Plaintiffs prevailed, the owners could have been exposed to unlimited liability for the injuries, cargo loss and above all, wrongful deaths.

If you multiplied "unlimited" by 1,500, the product would be more than enough to sink White Star [and, presumably, International Mercantile Marine]. As for our mythical complaint, it would only make sense to go forward if Mrs. Widener was trying to make a point, and wasn't concerned about attorney fees. It must have come as something of a shock to Madeleine Astor to discover that her husband's life was worth $92,000.

We have seen that the *Titanic* changed many things. One of them was the above law. The public and the press were furious over the meager liability limits available to ship owners in the United States based on post-casualty value of a vessel's remains, particularly in major tragedies such as the *Titanic*. As a result, amendments to the Law adopted a 'tonnage' based limitation fund similar to English Law. In loss of life and bodily injury cases, compensation increased to $60 per ton of the vessel in 1935 and in 1984, to $420 per ton. It was because of this amendment that Cochran got Mrs. Abbott her 1.5 million. To bad this law wasn't retroactive. Also swept away was the *"privity"* portion of the law, so from now on, simply put, if you own the boat, you're on the hook. Today, the idea that a lack of knowledge somehow mitigates guilt would be laughable.

Of course changing the law only guarantees some clever attorney will find a way to get around it. Ship owners now routinely limit liability by making each ship a separate LLC [Limited Liability Company] and leveraging them to the hilt, thereby rendering the ship virtually valueless. For what it's worth, others have learned this scam. If you are hit by a Yellow cab in New York and decide to sue, don't be surprised that your award of damages is anything more than the depreciated value of the cab. A "different" company owns each cab. [No, Shakespeare had it right, *"let's kill all the lawyers!"*]

Across the Atlantic, British courts wrestled with similar problems. There was, however, a difference. No Limited Liability Law protected English owners. The liability cap was pegged at two million dollars.

With so much more at stake, lawyers argued back and forth about the weather, the receipt of warning telegrams, and whether *Titanic* was proceeding at an unsafe speed. Each side had its own experts with clashing opinions about the nature of icebergs and the effects of haze and calm waters on their visibility. The jury deliberated briefly and found for the Plaintiffs on some grounds and against them on others. The shipping company appealed, arguing that the jury *"didn't know what it was doing."* [In case you are wondering, the concept of *non compos mentis* (not in your right mind) applies to defendants, not juries.] This battle continued through the courts, with the Defense winning enough times to effectively block big payouts to Plaintiffs. The victims never really got what they asked for [or deserved].

Nonetheless, if you were an American and had the means, the smart course would be to sue in a British Court. That would take the $92,000 out of play. The British High Courts, not wanting to be out done by the *"Supremes"* had a queer ruling of their own. Guess what an American lacks to sue a British Company, owned by Americans in a British Court? You guessed it: "standing."

For poor families lacking the means to wage a protracted legal battle, the end result on both sides of the Ocean was most often an out of court settlement. The largest settlement in the United States was five thousand dollars. In England, the biggest payout was fifty thousand and that to a corporation, not an individual. Insurance claims for property: clothes or jewels [or a new Renault] brought by primarily First Class passengers were not challenged and paid straightaway.

Chapter 16
Dénouement

*"Our inventions are wont to be pretty toys,
which distract our attention from serious things.
They are but improved means to an unimproved end."*

Henry David Thoreau

In the spirit of a James Patterson mystery, I've made you wait until the final page to find out *"who dun it?"* We have answered several questions, but now it is time to tackle the big ones. What ship did the *Californian* see, and what ship did the *Titanic* see? And did they see each other? This is the argument that has raged on since the day the *Californian* "came 'round" the tip of Cape Cod and docked in Boston. So, let's get to it.

In the world of *Titanic,* there are two kinds of people: The *Lordites* and The *Anti-Lordites*. The *Lordites* are the ones who believe that the *Californian* did not see the *Titanic* and that Captain Lord is blameless. On the other side, the *Anti-Lordites* are the ones who don't buy his story, and believe Lord is guilty as sin. Down through the years, each side had produced maps and charts and graphs and lord [no pun intended] knows what else to prove their claim. Each side can bring forth mathematicians and theorists to make their case.

Not to be outdone, I have my own theorist. Say hello to William of Occam, [a Franciscan friar I brought in from the 14th century]. He is the man who gave the world the principle known as *Occam's razor*. *Occam's razor* is the law of parsimony, economy or succinctness. It is the principle urging one to select from among competing hypotheses, that which makes the fewest assumptions. Reduced to its basics, it says [more or less] this:

> *"All things being equal, the simplest explanation is probably the correct one."*

It didn't take much ruminating to make my selection. There is no getting away from the basics. From the bridge of the *Californian*, officers saw a large steamer ablaze in light. None of the three ships she would most likely be mistaken for were anywhere nearby.

Her sister, *RMS Olympic*, was 500 miles away, outbound from New York. As for her rivals, *RMS Mauretania* [now with stoker Coffey] was off the Irish Coast. *RMS Lusitania* was in drydock at Liverpool.

At 11:40pm, the officer on watch observed the liner put out her lights. At 11:40 *Titanic* turned through 90 degrees, so her lights would not be visible from the *Californian*. There is much more evidence, but let's get to the dagger in the heart. The *Californian* observed the steamer fire rockets [eight rockets to be precise]. There is no proof that any ship, besides the *Titanic*, fired rockets in that area. How many did she fire? Eight. This is the simple explanation and the one I believe to be correct. In my mind, Mr. Occam got it right.

Over the century, new theories would rise and eventually fall. One example: Lord maintained that his ship was nineteen miles from the *Titanic*. The ship seen from the *Californian's* bridge was much closer. With the discovery of the wreck, any self-respecting *Lordite* would immediately jump up and say, *"Boxhall's position was wrong, so you can't say how far apart they were!"*

In the spirit of Occam, the answer is simple. The same instruments and techniques [sextants and star sights] that determined the *Titanic's* position were used to get a fix for the *Californian*. If one location is wrong, it is reasonable to assume the other probably is as well. Thus, the "position" argument doesn't stand muster. In the end, we keep coming back to the basics [and Occam].

My name is Steve Orlandella, and I am an unrepentant shipaholic. [The first step is admitting that one cannot control one's addiction.] I have, on numerous occasions visited the *SS United States* to pay my respects and have my heart broken at the sight of her slow decay while moored alongside Pier 82 in Philadelphia's Delaware River.

I have also made many a pilgrimage to the *RMS Queen Mary*. In her time she was the *"Queen of the Seas,"* as well known as any ship afloat. Now, she is spending her golden years still holding court in the warmth and sunshine of Long Beach, California.

She held the *Blue Ribband* throughout the War years and for seven more when she returned to peacetime service. Then in 1952, surrendered it to the current [and almost certainly the last] holder of the prize, the *SS United States*.

During one of those visits to Long Beach, I walked past a father and son standing on the quayside, studying her intently. The father said, *"This is a famous ship."* The little boy took a beat, and then asked, *"The most famous?"* I was out of range when the father replied, but it didn't matter, I knew the answer. The answer is, of course, no. In spite of over one thousand Atlantic crossings, speed titles, and a sterling war record, *"the Mary,"* and for that matter, the *Mauretania*, the *Olympic,* the *Normandie* and even the *United States,* will forever play second fiddle to a liner that didn't complete even a single voyage. She was, is, and forever will be, The Royal Mail Steamship *Titanic*.

Chapter 17
F. A. Q.

"I wish I had an answer to that because I'm tired of answering that question."

Yogi Berra

Did Nazi Germany make a movie about the *Titanic*?

Indeed they did, and a very expensive one at that. In 1940, the German minister of propaganda, Joseph Goebbels, commissioned a movie entitled *"Titanic."* It was an attack piece on the British, meant to show the Germans the kind of people they were *"up against."* The plot of the movie is centered on, who else? J. Bruce Ismay. Ismay is deliberately trying to drive down the price of White Star stock. He starts a rumor that the Line is in trouble because of the huge expenses in building the *Titanic*. His plan is to buy back the shares at the devalued price, and then announce from the *Titanic* that the ship is setting a record for the Atlantic crossing, thereby driving the price up again. The Nazis framed him as an evil British plutocrat scheming to profit off the backs of everyone else, especially the lower classes who had put their pitiful savings into the ship. The hero of the film was, of course, a German. An officer named Peterson warns of danger and ultimately saves a baby during the wreck, demonstrating to the audience all that is good, pure and selfless in Germans.

The Nazis lavished time and attention on the film with state-of-the-art effects. [I have seen the movie, and visually it is very impressive.] The film was finally ready in the spring of 1943, but by then, the German populace knew precisely who they were *"up against."* The country that was raining bombs on *"das Vaterland"* every night. In a rare moment of clarity, Goebbels decided not to release the film.

Unfortunately the story doesn't end there. The ship that "played" the *Titanic* [exteriors only] was the old HAPAG liner *Cap Arcona*. At war's end, she was used to transfer survivors of the *Neuengamme* concentration camp when she was sunk by mistake by RAF bombers. Five thousand souls, having survived the horrors of the *SS*, perished.

Titanic **(Germany, 1943)**

Harry Elkins Widener Memorial Library, Harvard

What is the connection between Harvard and the *Titanic*?

This is one urban legend that is absolutely true. Harry Elkins Widener, the son of George and Eleanor Widener, was lost along with his father when the *Titanic* sank. Harry was a Harvard graduate [Class of '07] with a passion for books. At the time of his death he was 27 and had a large and varied collection. Within a month after the disaster, Mrs. Widener approached her son's *alma mater* with a proposition. She was willing to donate $3,500,000 for a new main library and, with it, her son's book collection. The proposal came with two strings. First, the library must be named after her son. No problem. On June 24, 1915, The Harry Elkins Widener Memorial Library, located across from the church in Harvard Yard, was opened to the public. And what was the second string? Eleanor believed that her son died, not because the water was below freezing, but rather because he couldn't swim. She wanted every Harvard graduate to know how to swim. Thus, sometime during their senior year, every member of the Crimson must jump into the campus pool, swim across and swim back. The rule is rather loosely interpreted however, with more than a few undergrads using the dog paddle, nevertheless, *"a rule is a rule."*

Today *"the Widener"* serves as the centerpiece of the 15.6 million-volume Harvard University Library system, the largest university library system in the world and a book lover's dream come true.

Could the *Titanic* have actually been the *Olympic*?

This one requires some explaining. There was a man named Robin Gardiner [part-time historian and full-time plasterer] who wrote a book: *"Titanic-The ship that never sank."* In it, he put forth a massive conspiracy that went terribly wrong. Here we go.

Following the collision between the *Olympic* and *HMS Hawke,* the Liner was drydocked at Harland & Wolff. The Royal Navy Board of Investigation had exonerated the *Hawke* [imagine that], meaning that White Star would not collect on the insurance. The damage was far worse than expected, including damage to the keel. Putting it right would take months of work, and, since workers and parts would have to be diverted from the *Titanic,* the result would be financial disaster. The builders decided that it would be quicker to complete the *Titanic* than to repair the *Olympic.* For two months, the workers removed the promenade windows and the other changes on B-Deck and, for good measure reduced the size of the *á la carte* Restaurant and removed the *Café Parisien.* When the work was completed, she sailed for Southampton to begin her career as the *Olympic.*

Then a decision was made to dispose of the *Olympic* [now *Titanic*]. Superficial repairs were made and the Promenade Suites were installed. Now it gets interesting. The plan was to have the *Titanic* ram another ship in mid-ocean at a predetermined spot. Other ships [including the *Californian*] were pre-positioned to affect a rescue. They would dispose of the damaged ship, no lives would be lost, and insurance would help pay for a replacement. That was the game plan. As it turned out, it was *"game called on account of ice."* Believe it or not, this story had traction. Did it occur to anyone that a conspiracy of this size would involve thousands of construction workers at Harland & Wolff, the front office at White Star, the senior officers of the *Titanic,* as well as officers of other ships? [My guess is the secret would have survived until the first Irish shipwright got to a pub with his *"hush money."*] Sanity was restored when submersibles searching the wreck in 1990 found structural parts stamped "401." [The real *Titanic* was, if you recall, Harland & Wolff's hull No. 401.]

Was there a problem with the *Titanic's* construction?

Following Dr. Ballard's discovery, small pieces of the hull were brought up for analysis. Scientists demonstrated that the hull pieces had jagged edges, as if they had fractured like crystal. Their explanation was the high sulfur content of the steel. Then they went after the rivets. The rivets had too much slag and were prone to failure. While all this may be true, it still doesn't explain the cause of the damage. For that we need to enter the realm of Physics. I have worked with a Physics Professor at UCLA on this problem. The *Titanic's* gross weight was about fifty-two thousand tons, and we'll assume the *"immovable object"* weighed in at about two million tons. *Titanic* sideswiped the iceberg, but, at the same time, she was turning into the object. Murdoch put the helm over *"hard-a-starboard"* [turning left] in his attempt to get around the ice. The swing of the bow away from the berg turned her hull into the ice. Allowing for a turning speed of 10 knots, the deceleration to zero at the point of impact would generate 16×10^9 $^{KG}/H^2$, an enormous amount of force. [This also answers the question, *"who hit whom?"* She hit the berg].

Combining the twenty-five mile per hour "sideswipe" with the force pressing the ship against the iceberg should have resulted in far more damage, and yet it didn't. Her sister, the *Olympic*, was made of exactly the same steel, the same rivets and the same construction techniques. She survived 25 years on the Atlantic, three collisions, and a hit by a German torpedo, [a dud] and never came close to sinking. [For what its worth, she also rammed and sank a *U-boat*.] They didn't call her *"Old Reliable"* because her steel was suspect. As for the rivets, the scientists tested exactly two. This would be two out of the three million that went into her hull and superstructure [you do the math]. These theories have fallen out of favor in recent years.

Did they recover Wallace Hartley's violin?

As the revision of this book goes to press in August 2013, the issue [for most, but not all] seems to have been settled. Many believe that Hartley's violin was lost in the wreck. A violin was found in the attic of an English house, inside a leather case with the initials *"W. H. H."* [Wallace Henry Hartley]. The dispute lies in the conflicting stories about the recovery of Hartley's body. Some reports said that the violin was strapped to his waist, while others claimed it was his music case [containing just sheet music]. Story "A" says that the violin was recovered from the corpse and shipped to his grieving fiancé. Story "B" claims that when the case was opened inside was nothing but his music, including the charts for *"Nearer my God to thee."* "B" also asserts that the detailed list of items recovered from Hartley's body [including the coins in his pockets] makes no mention of the violin or the case. Corroboration for story "A" came from an inscription on the instrument and [what is purported to be] Maria Robinson's diary,

> *"I would be most grateful if you could convey my heartfelt thanks to all who have made possible the return of my late fiancé's violin."*

Story "A" seems to have won out, especially after nearly two years of forensic testing determined that both the violin and case were "genuine." Regardless, the *Titanic Historical Society* remains unconvinced, *"Every February or March as the anniversary nears, there is an article, a supposed Titanic artifact found."* Others will tell you that a professional quality violin, such as the one found would be beyond the means of a *"café violinist"* like Wallace Hartley. As for the forensic testing - keep in mind it was paid for by people who had a vested interest in the result, the owners and auctioneers. [Me? I side with the *"THS."* It is not Hartley's violin. It is a hoax.]

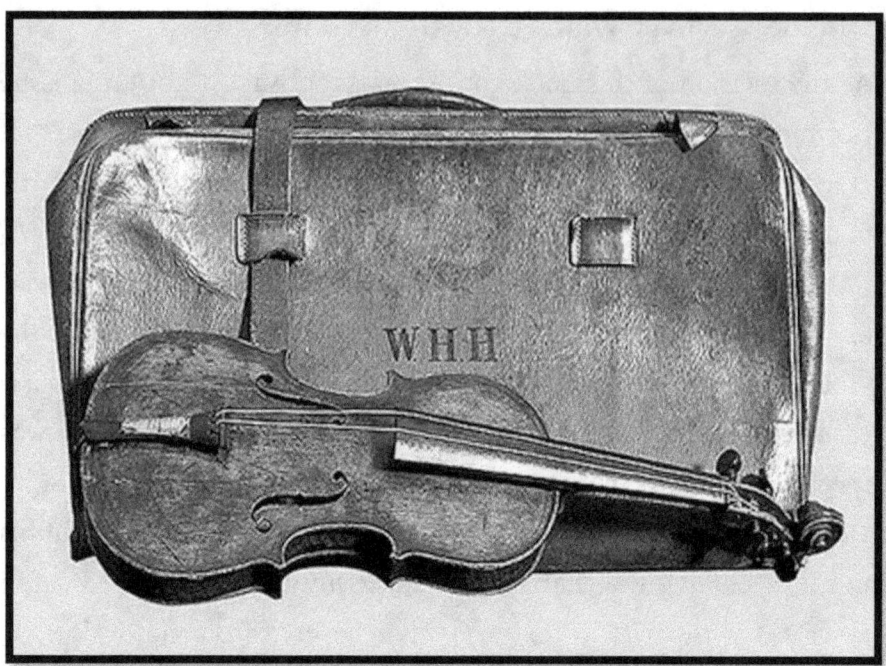

Purported to be Wallace Hartley's violin & case

NO. 224.

MALE.   ESTIMATED AGE, 25.   HAIR, BROWN.

CLOTHING—Uniform (green facing); brown overcoat; black boots; green socks.

EFFECTS—Gold fountain pen, "W. H. H."; diamond solitaire ring; silver cigarette case; letters; silver match box, marked "To W. H. H., from Collingson's staff, Leeds"; telegram to Hotley, Bandmaster "Titanic"; nickel watch; gold chain; gold cigar holder; stud; scissors; 16s.; 16 cents; coins.

BANDMASTER WALLACE H. HOTLEY.

Permit issued May 3, 1912.
Liverpool, England.

Items found on the body of Wallace Hartley (misspelled).

Did a man survive by dressing as a woman?

This is one of the ship's most enduring urban legends, which I can refute with an emphatic...not really. A man dressing up as a woman to escape was mostly just that, a legend.

There are several possible suspects. One was Daniel Buckley, a 21-year-old Third Class passenger from Ireland who climbed into boat No. 13 with a mixed group, including other male passengers. He later said that Mrs. Astor saw him crying and threw her shawl over him.

Fifth Officer Lowe saw somebody suspiciously dressed like a woman get into boat No. 14. This may have been Edward Ryan, an Irishman from Third Class. Ryan's published account stated that he saved himself and a woman by climbing down a rope into a lifeboat. But he wrote to his parents:

> "I had a towel round my neck...I wore my waterproof overcoat. I then walked very stiff past the officers...they didn't notice me. They thought I was a woman."

The man most victimized was the aforementioned William T. Sloper. Dorothy Gibson had already entered Lifeboat No. 7 and insisted that he join her. First Officer Murdoch, who had been allowing men into the early boats, gave him permission to board. After the *Carpathia* reached New York, a *New York Herald* reporter fabricated a story, which identified him as having dressed in women's clothing to escape the ship. On advise from family, he chose not to sue the *Herald* or the hack reporter. He reasoned that the fuss would, in time dissipate. It did not. Sloper would spend the rest of his life refuting the charge.

And so the legend endures. The great character actor, Allyn Joslyn, portrayed the threadbare First Class interloper *"Mister Meeker"* to perfection in the 1953 film version. In a lifeboat, *"Maude Young"* [Thelma Ritter's loosely drawn characterization of "Maggie" Brown] spots a pair of men's shoes sticking out from under a dress. She pulls off the person's hat and remarks, *"I see you made it, Mister Meeker."*

Why did "hard a-starboard" turn the ship to the left?

This is usually the first question asked by a sailor. Before modern standardization, British Quartermasters were advised to follow the rotation of the bottom of the wheel. Thus, when obeying a "hard a-starboard" command, the QM would turn the bottom of the wheel to the right, or starboard. This applied left rudder and would turn the ship to the left [to port]. Steering with the bottom of the wheel was an approved way to learn helming more than a century ago. The nautical reason for a "hard a-starboard" command to turn left seems related to the tiller and not the rudder. The tiller is pushed to the right, or starboard, to apply left rudder and turn the vessel to port.

With enough lifeboats would everyone have survived?

Probably not, since the most critical element was not lifeboat space but time. In the two hours and forty minutes between her striking the berg and the time she foundered, the crew managed [barely] to launch eighteen boats [the Atlantic *"launched"* the other two] filled to 60% of capacity. Given their situation more boats would have, in all likelihood, just cluttered the Boat Deck. Nevertheless, the combination of a well-drilled crew, more and better positioned boats, use of the power hoists, a more decisive Captain, the prompt evacuation of Third Class passengers, and above all, a heightened *"sense of urgency"* would have lowered the death toll dramatically. The reluctance of passengers to abandon a ship many believed to be "unsinkable," meant certain death, *"for those in peril on the sea."*

Was First Officer Murdoch drunk?

Entwined with legend of the *Titanic* are stories about liquor and intoxication. The most recent surfaced five years ago in a book entitled: *"The Man Who Sank Titanic."* The book was written by Sally Nilsson and is, in fact the biography of her grandfather, Quartermaster Robert Hichens. [He of Lifeboat No.6 fame.] So, who gets the bame? It falls on the First Officer, William Murdoch. Why? Because Murdoch, the officer on watch, was drunk [on wine] and passed out on the bridge. All this was learned in a letter that surfaced some time ago. The letter is reprinted on Page 378. The story is told by someone [Hichens] to a man named Henry Blum, who then told it to a Thomas Garvey. Is it true? Let's pick it apart, shall we?

The letter is problematic from the start because it's triple-hearsay. No court in the world would accept a document like this and neither should we. Hichens is presumed to be the Harbormaster, a job he was completely unqualified to do. His life is well known and there is no record of him ever being in Cape Town. There was no "lounge" anywhere on the bridge, in fact, there was no place to even sit down. A man whose job was to work on the bridge [a Quartermaster] would know that. He then claims he heard the crow's nest lookout shout "Iceberg, dead ahead!" The crow's nest was over a hundred feet away. Messages went to the bridge via telephone. Also, the correct message would have been, "Iceberg, RIGHT ahead!" He goes on to say he heard the bow lookout say rhe same thing. The *Titanic* had no bow lookout. The book alleges that Murdoch had become intoxicated at a celebration for the Captain. The only "celebration" for Smith was the Widener's dinner party in the *à la carte* restaurant. There is no evidence of Murdock even being in the room, let alone downing vino. Is there a possibility that this story is true? No, there is not.

To Whom it may concern:

Regarding the S.S. Titanic.

While serving as quartermaster on a British ship, they called at Capetown in 1914 - The Harbormaster came aboard and after pledging Henry Blum to secrecy for 10 years, related the following account saying he wanted to tell someone to relieve his conscience.

He had been, he said, "the quartermaster on the Titantic doing his 2 hour trick at the wheel the night of the disaster. He heard the Look-out in the crows nest call out,'Iceberg dead ahead.' Seconds later the Bow Look-out called out 'Iceberg dead ahead'. The First Officer was lying on the lounge at the rear of the pilot house. The Quartermaster said he then shouted the warning in the First Officer's ear but could not awaken him. He then returned to the wheel and held her steady on her course as he should'. When the survivors were rounded up he was placed under house arrest and spirited away to Capetown. He was given a life long job with good pay for as long as he remained silent."

                                      Thomas Garvey

Reprint of the Garvey letter.

Chapter 18
Destiny

"Like as the waves make towards the pebbl'd shore, so do our minutes, hasten to their end."

William Shakespeare

Madeleine Force Astor came off the *Carpathia* and went into seclusion. In August 1912, she gave birth to John Jacob Astor VI at her Fifth Avenue mansion. For the next four years, she raised him as part of the Astor family. Madeleine's second marriage was to her childhood friend, banker William Karl Dick, Vice President of the Manufacturers Trust Company of New York. The couple had two sons together but got a Reno divorce in July of 1933. Her third and final trip down the aisle came just four months later when she married a twenty-six-year-old Italian boxer named Enzo Fiermonte in a civil ceremony in New York City. They divorced five years later, and she resumed using the surname of her second husband. Madeleine Dick [*née* Madeleine Force Astor Fiermonte] died of a heart ailment in Palm Beach, Florida, in March 1940, at the age of 46.

RMS *Aquitania* sailed on her maiden voyage on May 30, 1914. She had completed three round trips before the assassination of Archduke Franz Ferdinand plunged the world into war. As with her sisters, she was unsuited for the role of armed merchant cruiser. After being laid up for a time, she was converted into a trooper in the spring of 1915, and made voyages to the Dardanelles, often times running alongside the *Britannic* or the *Mauretania*. During this period she also served as a hospital ship. Following the war, she became one of the most popular liners on the *"Atlantic Ferry."* The *Aquitania* was scheduled to be replaced by the *RMS Queen Elizabeth* in 1940, but that plan was shattered with the coming of the Second World War. Once again she was used as a troopship, becoming the only liner to serve her country in both World Wars. By the end of the conflict her condition had deteriorated, and she was retired in 1950. Later that same year, the last four-funneled passenger ship went to the breakers in Scotland.

David "Davy" Blair left the *Titanic* on 9 April 1912, taking with him the key to the crow's nest locker, probably by accident. This is believed to be the reason why there were no binoculars available to the crew during the voyage. At the Senate Inquiry, lookout Frederick Fleet was asked if binoculars would have made a difference. He replied, *"enough to get out of the way."* Blair was First Officer on the *Majestic* in 1913, when a stoker jumped overboard the night after a fellow crewmember had succeeded in drowning himself. While a lifeboat was organized, he jumped into the ocean and swam toward the man who was now swimming for the ship. Though the boat reached the man first, Blair was commended for his bravery and received money from the passengers and a medal from the Royal Humane Society. He served aboard the *RMS Oceanic* when it ran aground in 1914. As the navigator, he received the blame for the grounding at the resulting court martial. It is believed David Blair perished in 1917 when his ship was a torpedoed by a German U-Boat in the Atlantic. The key [its origin still the subject of debate] survived and was donated by Blair's daughter to the International Sailors Society. In 2007, *Christie's* auctioned it off. A jeweler purchased it. The Price? £90,000. [$130,000]

Joseph Groves Boxhall briefly served as Fourth Officer on White Star's *Adriatic* before joining the Royal Naval Reserve as a sub-lieutenant. During the First World War, he served on a battleship and commanded a torpedo boat. Boxhall returned to White Star following the war in May 1919. He signed on to serve as Second Officer aboard *RMS Olympic* in June 1926. After the White Star-Cunard merger in 1933, he served as First and later, as Chief Officer of the *RMS Aquitania,* although he never made captain in the merchant service. After 41 years at sea, he retired in 1940.

The last surviving deck officer of the *Titanic*, Boxhall died of a cerebral thrombosis on 25 April 1967 at the age of 83. According to his last wishes, his ashes were scattered at 41°46N, 50°14W, the position he had calculated [and defended] as *Titanic's* final resting place fifty-five years earlier.

Margaret "Maggie" Brown used her new fame as a platform to talk about issues that deeply concerned her: labor rights, women's rights, education and literacy for children, and historic preservation. During World War I, she worked with the *American Committee for Devastated France* to help rebuild areas behind the front line and worked with wounded French and American soldiers at a museum outside of Paris. In 1932, she was awarded the French *Legion of Honor* for her "overall good citizenship," which included helping organize the *Alliance Française,* her ongoing work in raising funds for *Titanic* victims and crew, her work with Judge Ben Lindsey on the Juvenile Court of Denver, and her relief efforts during World War I. At the age of 65, a brain tumor ended her life on October 26, 1932. On November 3, 1960 a new musical opened at the Winter Garden Theatre on Broadway. The Title was *"The Unsinkable Molly Brown."*

SS Californian continued in commercial service until the outbreak of World War I, when she was taken over by the British Admiralty. On November 9, 1915, while en route from Salonika to Marseille, she was torpedoed and sank approximately 61 miles southwest of Cape Matapan, Greece, by the Imperial German Navy submarine *U-35,* with the loss of one life. The wreck has yet to be found.

Helen Churchill Candee subsequently gave a short interview about her experiences to the Washington Herald and wrote a detailed article on the disaster for *Collier's Weekly.*

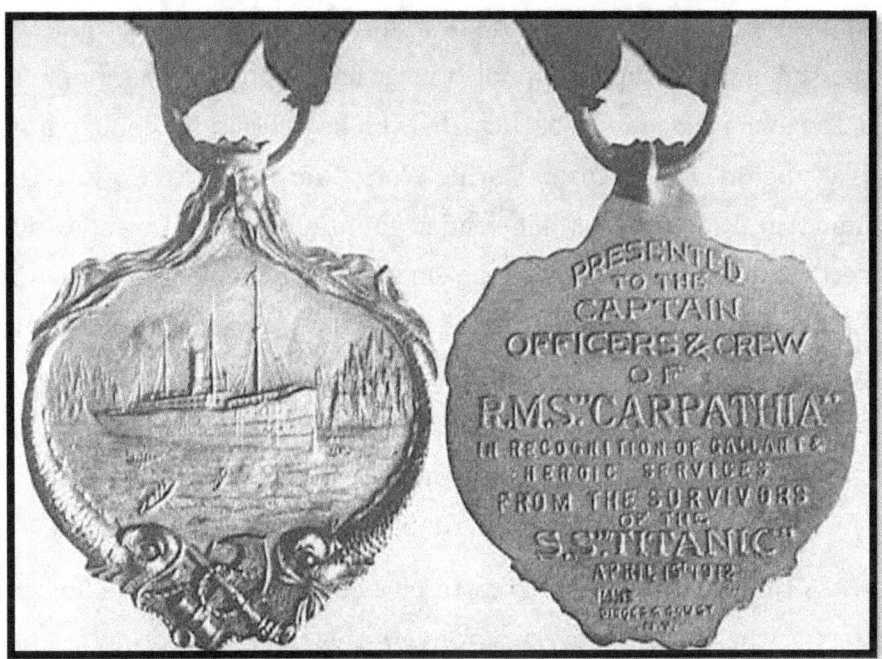
Medal awarded to Captain and Crew, *RMS Carpathia*

Captain Arthur Rostron and Margaret "Maggie" Brown

The cover story was one of the first in-depth eyewitness accounts of the sinking published in a major magazine. Active in the Women's suffrage movement, she participated [on horseback] in the march on Washington D. C. in 1913. During World War I, Candee worked as a nurse in Rome and Milan under the Italian Red Cross, which decorated her for her service. One of her patients in Milan was Ernest Hemingway. At the age of 80, Candee was still traveling and writing. The *Mackay-Bennett* had found the body [#258] of a *"gentleman in a grey suit."* Among his effects was a cameo of a woman. It was returned to Mrs. Candee. Helen died in 1949, at her summer cottage in York Harbor, Maine. She was ninety years old.

RMS Carpathia and her crew received numerous awards following their rescue of the *Titanic* survivors. During the First World War, the *Carpathia* was used to transfer Canadian and American troops to Europe. She served as a troopship by the Canadian Expeditionary Force, sailing from New York through Halifax to Liverpool and Glasgow. On 15 July 1918, she departed Liverpool in convoy. On the morning of July 17, she was torpedoed, at 9:15am, in the Celtic Sea by the Imperial German Navy submarine *U-55*. All 57 passengers (36 Salon Class and 21 steerage) and 218 surviving crewmembers made for the lifeboats as the vessel settled by the head and were rescued. *RMS Carpathia* sank at 11:00am at position 49" 25' N, 10" 25' W, approximately 120 miles west of Fastnet.

Collapsible "A" was found a month after the disaster, on 13th May 1912. The White Star Line ship *RMS Oceanic* spotted the lifeboat, and the three dead bodies still aboard. The bodies in the boat were believed to be that of a fireman, a sailor, and First Class passenger named Thomson Beattie. They were buried at sea from the lifeboat. "A" was taken aboard Oceanic, after which its fate is unknown.

RMS Gigantic was a name without a ship. Conceived as the third *Olympic*-class liner, it is believed her name was changed to the less ostentatious *RMS Britannic* before [and not after] the loss of the *Titanic*. Launched in February 1914, she finally entered service as a hospital ship [*HMHS Britannic*] in December of the following year. She was used to transport wounded back from the disastrous Gallipoli campaign. In that role, she struck an underwater mine off the Greek island of Kea in the Kea Channel [Aegean Sea] on the morning of November 21, 1916, and sank with the loss of 30 lives. The Bob Hope movie, *"The Big Broadcast of 1938,"* was centered on a transatlantic race between two massive ocean liners. The loser was named the *Colossal*; the winning ship was [wait for it] the *Gigantic*.

Harland & Wolff persevered in spite of the loss of its leader. During World Wars I & II, the shipyard built 140 ships for the Royal Navy - including *HMS Belfast*. In the 1950's, the yard concentrated on overhaul, re-fitting and ship repair. On February 1, 2011, it was announced that Harland & Wolff had won the contract to refurbish the tender *SS Nomadic*. The *Nomadic* was laid down in the Queen's Island shipyard in 1910. [Hull No. 422] Designed by Thomas Andrews, she was built on slipway No. 1 alongside *RMS Olympic and RMS Titanic*. The *Nomadic* is the last surviving White Star vessel.

Eva Hart remained active in *Titanic*-related activities well into her 80s. In 1982, she returned to the United States and joined several other survivors at a Titanic Historical Society convention commemorating the 70th anniversary of the *Titanic* sinking. In 1994, Hart wrote an autobiography, *"Shadow of the Titanic – A Survivor's Story,"* in which she described her experiences aboard the ship and the lasting implications of her sinking. She openly criticized the

White Star Line. She was also critical of salvage efforts, labeling the salvors, *"fortune hunters, vultures, pirates, and grave robbers."* Hart died on February 14, 1996 at her home in Chadwell Heath at the age of 91. At the time, her death left eight remaining survivors.

William Randolph Hearst saw his news empire reach a circulation and revenue peak in 1928, but the economic collapse of the Great Depression and the vast over-extension of his empire cost him control of his holdings. The newspapers never paid their own way. Mining, ranching and forestry provided the dividends the Hearst Corporation paid out. When the collapse came, all the Hearst properties were hit hard, but none more so than the papers. The Hearst Corporation faced a court-mandated reorganization in 1937. Newspapers and other properties were eventually liquidated, and in 1947 he left *"La Cuesta Encantada,"* his opulent castle in San Simeon, California. Four years later he died of a heart attack in Beverly Hills. He was eighty-eight.

Ben Hecht served as war correspondent in Berlin for *The Chicago Daily News*. In New York, he co-wrote on a play, *"The Front Page"* [by any standard, a classic American comedy]. He would become one of the hottest screenwriters in the business, earning him the nickname, *"The Shakespeare of Hollywood."* He died April 18, 1964.

Robert Hichens testified before both the Senate and Board of Trade inquiries. In World War I, Hichens served in the Royal Naval Reserve. In the 1920's, he moved his family to Devon. He started a boat charter business, but ran up big debts and lost the boat. In 1931, his wife left him and moved to Southampton. An alcoholic, he spent the next year looking for work, without success. After an incident with a gun, he was arrested and attemped suicide. After prison, he served aboard the cargo ship *English Trader*. He died at sea in 1940.

Masabumi Hosono survived but found himself ostracized by the Japanese public, press and government for his decision to save himself rather than die with the ship. It was said he had *"betrayed the Samurai spirit of self-sacrifice."* Hosono soon found himself thetarget of public condemnation. He lost his job and was condemned as a coward by the Japanese press. He died in 1939.

Joseph Bruce Ismay never really recovered from the loss of the *Titanic*. Ismay stayed out of the public eye for most of the remainder of his life. He retired from active affairs in the mid-1920s, and lived with his wife on a large estate in Ireland. His health declined in the 1930s following a diagnosis of diabetes, which took a turn for the worse in 1936. The illness resulted in the amputation of part of his right leg. He returned to England a few months later, settling in a house across the River Mersey from Liverpool, the city his ships abandoned in 1907. He died in London [Mayfair] October 17, 1937, just steps away from No. 24 Belgrave Square, where he and Pirrie first envisioned the *Titanic*. *The New York Times* eulogized him as:

> *"A man of striking personality and in any company arrested attention and dominated the scene. Those who knew him slightly found his personality overpowering and in consequence imagined him too be hard, but his friends knew this was but the outward veneer of a shy and highly sensitive nature, beneath which was hidden a depth of affection and understanding which is given to but few. Perhaps his outstanding characteristic was his deep feeling and sympathy for the 'underdog' and he was always anxious to help anyone in trouble. Another notable trait was an intense dislike of publicity, which he would go to great lengths to avoid."*

Charles Herbert Lightoller continued with the White Star Line. He was decorated for gallantry as a naval officer in the First World War [and shipwrecked again]. He returned to White Star but soon learned that opportunities for advancement within the line were no longer available to anyone who served on the *Titanic*. Lightoller resigned shortly thereafter, taking such odd jobs as an innkeeper and a chicken farmer and later, property speculator. He would serve his country one more time in the Second World War by providing and sailing one of the "little ships" during the evacuation of Dunkirk. Charles Lightoller died December 8, 1952, of chronic heart disease. In *"A Night to Remember,"* Walter Lord would chronicle his heroism.

Stanley Lord was dismissed by the Leyland Line in August of 1912. The Captain, the *Californian's* officers and crew having been *"damned with faint praise."* The conclusions of both the United States and the British Inquiries seemed to disapprove of the actions of Captain Lord but stopped short of recommending charges. In February 1913 the Nitrate Producers Steamship Company hired him, where he remained until March 1927, resigning for health reasons. In 1955, following the release of Walter Lord's [no relation] bestseller, *"A Night to Remember"* and the subsequent film of the same name, Lord, embarrassed at his portrayal in the movie, attempted to clear his name. Two separate petitions to the Board of Trade were rejected. Captain Lord died on 24 January 1962.

Harold Godfrey Lowe returned to his hometown of Barmouth, to a reception held in his honor at the Picture Pavilion. He was presented with a gold watch, with the inscription *"Presented to Harold Godfrey Lowe, Fifth Officer RMS Titanic by his friends in Barmouth and elsewhere in recognition and appreciation of his*

NOTICE!

TRAVELLERS intending to embark on the Atlantic voyage are reminded that a state of war exists between Germany and her allies and Great Britian and her allies; that the zone of war includes the waters adjacent to the British Isles; that, in accordance with formal notice given by the Imperial German Government, vessels flying the flag of Great Britian, or of any of her allies, are liable to destruction in those waters and that travellers sailing in the war zone on ships of Great Britian or her allies do so at their own risk.

IMPERIAL GERMAN EMBASSY,
WASHINGTON, D. C., APRIL 22, 1915.

Notice placed in *The New York Times*, May 1, 1915

gallant services at the foundering of the Titanic, 15th April 1912." He twice apologized to the Italian Government for using the word "Italian" as a sort of synonym for "coward." He served in the Royal Navy Reserve during the First World War, attaining the rank of Commander before retiring to Degenwy, Wales, with his wife and family. He died of hypertension on 12 May 1944 at the age of 61.

RMS Lusitania remained in regular service on the Atlantic run, even after The First World began in August 1914. The Admiralty chose not to use her in the role she was built for. M'Lords reasoned the amount of coal she consumed could be put to better use in the boiler rooms of a battle cruiser [or perhaps, eight destroyers].

The *"Lucy"* carried on, even though bookings were low. On May 1 1915, she left New York bound for Liverpool. To save coal, her speed was cut to twenty-one knots; this despite a warning from the German Embassy in the newspapers that the waters around Britain were now *"a war zone,"* and, ships would be *"liable to destruction."* On May 7th, off the Irish coast, abeam of the *Old Head of Kinsale,* she took a direct hit from a torpedo fired by the submarine *U-20*. Something [coal dust or more likely, artillery shells and gun cotton] triggered a second explosion that ripped open her hull. She settled so quickly that only six of her forty-eight boats could be lowered. In eighteen minutes, Cunard's safety record was gone and with it 1,198 souls.

RMS Mauretania would become Cunard's *"Golden Ship."* In 1914, at the opening of hostilities, she was acquired by His Majesty's Government for use as an armed merchant cruiser, but her huge size and massive fuel consumption made the liner unsuitable for the duty. [As with the others, she couldn't do the job she was built to do.] The Liner was *"laid up"* [held in reserve] in Liverpool until August 1915.

HMT (RMS) Olympic in her "war paint," Gallipoli, 1915

HMHS Britannic (Formally *RMS Gigantic*)

The *"Maurey"* was about to fill the void left by the loss of her sister, the *Lusitania*, when the Lords of the Admiralty ordered her to serve as a troopship during the Gallipoli Campaign. She reached her zenith in the post-war decade under Captain Rostron, becoming known as the *"Rostron Express"* for her speed and on-time performance. In 1928 she surrendered the *Blue Ribband* to the Germans and in September 1934 she was taken out of service. A new *"superliner"* was building on the Clyde. The *RMS Queen Mary* had rendered her superfluous. In July of 1935, she went to the breakers at Jarrow, one bend of the Tyne away from her birthplace. Her demise was protested by many of her loyal passengers, including President Franklin D. Roosevelt. Fifteen years later an American company would use U. S. Navy money to build the fastest liner of all-time, the *United States*. [What was that about history repeating itself?]

RMS Olympic enjoyed the long and glorious career cruelly denied to her sisters. In October 1912, she returned to Harland & Wolff for a *"re-fit."* Raised bulkheads, an extended "double bottom" and more pumps meant that she could survive a collision like the one that claimed her sibling. In May 1915, [now billed as the *new Olympic*] she was "drafted" by the Admiralty. HMT [His Majesty's Troopship] *Olympic* rammed the German submarine *U-103*, becoming the only merchant vessel to sink an enemy warship in World War I. She was very popular in the 1920's with many of her passengers wanting to sail on the *Titanic's* twin. The *Olympic* was withdrawn from service in April 1935, and after being laid-up for several months, was towed to the breakers at Jarrow for demolition, following a public auction. [Her paneling still graces the walls of many an English pub.] By her retirement, she had completed 257 round trips across the Atlantic transporting 430,000 passengers and covering 1.8 million miles.

Had she been spared until the Second World War, a ship of her size and capability would have been, according to many naval experts, *"worth her weight in gold."*

Herbert John Pitman remained with The White Star Line. In July 1912, he re-joined his old ship the *Oceanic* as her Third Officer and later served on the *Olympic*, although by then he had transferred to the Purser's Section because his eyesight was deteriorating. In the early 1920's, he moved from White Star Line to Shaw, Savill & Albion Company Ltd. During The Second World War, he served aboard *SS Mataroa* again as Purser. In all, he served the Company for 20 years. In 1946, just prior to leaving the Merchant Service, he was awarded an MBE [Most Excellent Order of the British Empire] for *"long and meritorious service at sea and in dangerous waters during the war."* *Titanic's* Third Officer died on December 7, 1961, he was 84.

Arthur Henry Rostron saw his star continue to soar after his gallant rescue of the *Titanic* survivors. He was awarded the *Congressional Gold Medal* by the United States Congress and was appointed *Knight Commander of the Order of the British Empire.* He quickly rose up the Line's Captain list, ultimately given command of *RMS Mauretania* and was made the Commodore of the Cunard fleet before retiring in 1931. When the *Mauretania* sailed for the breakers in 1935, Rostron was supposed to have been on board. Overcome with emotion, he refused to board her and instead waved farewell from pierside. He wanted to remember his beloved liner, as she was when he commanded her. Sir Arthur Rostron died of pneumonia in 1940 and is buried at the West End Church in Southampton. In his memory a plaque was placed in the New York Hall of Fame, an honor never previously accorded an Englishman.

Turbinia continued to demonstrate Parson's remarkable invention. In 1900, she steamed to Paris and was shown to French officials and then displayed at the Paris Exhibition. Seven years later she was on hand at Wallsend for the first sailing of *RMS Mauretania*. A decade later, she was officially retired [and left to rot]. In 1983, a complete reconstruction was undertaken. In October 1994, a century after her launch, *Turbinia* was moved to the Newcastle Museum. Then in 2000, she was added to the *National Historic Fleet, Core Collection*, thereby assuring her future. The little steam yacht now stands alongside *HMS Victory* and the other iconic ships of British History.

Carr Van Anda stayed as Managing Editor of the *New York Times*. The newspaper continued to prosper under Van Anda's management, and, by 1921, circulation had reached 330,000 during the week and 500,000 on Sunday. Advertising would increase tenfold in his twenty-five years at the helm. In the '20s it was said of him he was, *"the most illustrious unknown man in America."* A true *Renaissance Man*, he had an interest in archaeology and secured near-exclusive coverage of the opening of *Tutankhamen's* tomb by Howard Carter in 1923. He also famously corrected a mathematical error in a speech given by Albert Einstein that was printed in the *Times*. He retired from *The New York Times* in 1932. Upon learning of his daughter's death, he suffered a heart attack and died. He was 80 years old. The Scripps School of Journalism at Ohio University annually awards the "Carr Van Anda Award" to honor outstanding work by journalists.

The White Star Line's run of bad luck continued into The First World War, with the loss of the *Oceanic* and the *Britannic*. In 1922 the Line gained two ships, *Majestic* and *Homeric*, which had been seeded to England by Germany as War reparations. By 1933, White Star and Cunard were both in serious financial difficulties because of

The Great Depression, and the advanced age of their fleets. The British government agreed to provide assistance to the two competitors on the condition that they merge their North Atlantic operations. The agreement was completed on 30 December 1933. In 1947, Cunard acquired the remaining 38% of Cunard White Star, and on 31 December 1949 it acquired White Star's assets and operations. Thomas Ismay's company, The White Star Line, had ceased to exist.

Edward Smith, Henry Wilde, William Murdoch, and James Moody, the *Titanic's* three most Senior Officers, and the Sixth Officer, would not survive. Captain Smith, after leaving the Marconi cabin, returned to the bridge. In Lightoller's account he shut himself inside the wheelhouse and drowned clutching the ship's wheel. [Unlikely, since *"Lights"* was one deck below, grappling with "B."] The story of Smith swimming in the water with a baby and telling the crew of a lifeboat to *"be British"* is just that, a story. Where it came from, no one can say. [If I had to bet, my money would be on some Fleet Street editor who decided, *"Captain Smith must be a hero!"*] More is known about the First Officer. Murdoch was working to get the collapsibles away as the sea washed over the Boat Deck. Harold Bride said he saw him in the water a few minutes later, but he was already dead. Less is known about the Chief Officer. Wilde was last seen also helping with the collapsibles. A few survivors claimed he committed suicide. While this is possible, perhaps because his wife and children had died in 1910 and he now saw nothing left to live for, I still don't buy it. He was a career British Officer, sworn to do his duty, and, at that moment, his duty was helping to save lives, not reaching for a revolver. By rights Sixth Officer James Moody, like the other three Junior Officers, should have been in charge of a boat. As you recall, while loading No. 14, Fifth Officer Lowe remarked that an

Frank Beken's Photo, Solent Roades, April 10, 1912

41' 43" N, 49' 56" W, *RMS Titanic*

officer should man the lifeboat. The lower-ranked Moody would traditionally have been given this task, but he deferred to Lowe, and then moved forward to help Murdoch with the collapsibles. As the sea reached the bridge he dove into the water. His body was seen near Murdoch's with a head wound. Harold Bride theorized that he had been shot. As with Wilde, there is little to support the claim.

RMS Titanic lies at 41' 43" N, 49' 56" W., in 12,415 feet of water, that position is 13.5 miles away from Boxhall's 41' 46" N, 50' 14" W. The point where she sank was, in all likelihood, some distance away. The difference resulting from planing action through water. In total darkness, she faces north and sits upright on the bottom in two sections. One hundred years of immersion in the North Atlantic has done the job. The hull is slowly but inexorably decaying. Talk about raising her began even before the *Carpathia* reached New York. Someone put forth the absurd notion that people may still be alive, trapped inside the hull. In true *"Gilded Age"* spirit, plans were drawn up. Through the years all manner of schemes have been put forth, and that's all they were...schemes. The splendid fiction writer Clive Cussler, in his novel *"Raise the Titanic!"* has the Navy recover the wreck and then [speaking for every *Titanic* enthusiast in the world] towed triumphantly beneath *"Liberty"* and into New York Harbor.

But alas, that will never happen. Metal-eating bacteria has weakened the hull and bulkheads to the point where an attempt would not be feasible. Marine archeologists believe that sometime during the 21st century, the decay will reach the point where the hull will collapse upon itself, like a house of cards. Many of the items in the debris field, including bottles of [French] wine and her three [bronze] propellers will lie unchanged, but the legendary *RMS Titanic* will be little more than a patch of rust on the bottom of the North Atlantic.

Bibliography

*"I find television very educating.
Every time somebody turns on the set,
I go into the other room and read a book."*

Groucho Marx

Books

Ballard, Dr. Robert, *Finding the Titanic,* Cartwheel Books, 1993
Ballard, Dr. Robert, *The Discovery of the Titanic*, Warner, 1987
Beesley, Lawrence, *The Loss of the Titanic: Its Story & Its Lessons*
Bonsall, Thomas, *Titanic,* Gallery, 1987
Booth, John & Coughlan, Sean, *Titanic: Signals of Disaster,* 1993
Boyd-Smith, Peter, *Titanic: From Rare Historical Reports,* 1992
Brown, Rustie, *The Titanic, the Psychic, and the Sea,* Blue Harbor
Bullock, Shan, *A Titanic Hero, Thomas Andrews, Shipbuilder, 1914*
Butler, Daniel Allen, *Unsinkable, Full Story of RMS Titanic,* 1998
Butler, Daniel Allen, *The Age of Cunard,* Prostar
Cohen, Leo, *Titanic Revisited,* L. Cohen, 1984
Cooper, G.J., *Titanic Captain: The Life of Edward John Smith,* 2011
Costello, Philip, *Titanic*, Titanic Productions, 1985
Cussler, Clive, *Raise the Titanic!* Viking Press, 1976
Davie, Michael, *Titanic: Death and Life of a Legend,* Knoph 1987
Eaton, John P. & Haas, *Titanic: Destination Disaster,* Norton 1987
Eaton, John P. & Haas, *Titanic: Triumph and Tragedy*, Norton 1986
Gibbs, Philip, *The Deathless Story of the Titanic,* Lloyd's, 1912
Gracie, Archibald, *The Truth about the Titanic,* Kennerly, 1913
Harrison, Leslie, *A Titanic Myth, The Californian*
Harrison, Leslie, *Defending Captain Lord,* Image, 1996
Hart, Eva, *Shadow of the Titanic – A Survivor's Story,* 1997
Lightoller, Charles, *Titanic and Other Ships,* Nicholson-Watson, 1935
Lord, Walter, *A Night to Remember,* Holt Rinehart & Winston, 1955
Lord, Walter, *The Night Lives On,* Hyperion, 1992
Marcus, Geoffrey, *The Maiden Voyage,* Viking Press, 1969
Maxtone-Graham, John, *The Only Way to Cross,* Barnes & Noble
Neil, Henry, *Wreck and Sinking of the Titanic,* Homewood Press

Padfield, Peter, *The Titanic and the Californian,* John Day, 1965

Robertson, Morgan, *Futility: The Wreck of the Titan,* 1898

Rostron, Captain Arthur, *Home from the Sea,* Macmillan, 1931

Spignesi, Stephen J. *The Complete Titanic,* Birch Lane Press, 1998

Stenson, Patrick, *"Lights," The Odyssey of C. H. Lightoller,* 1984

Thayer, John B. *The Sinking of the S. S. Titanic,* Thayer, 1940

Wade, Wyn Craig, *Titanic: End of a Dream*, Penguin, 1986

Newspapers

The Boston American, April 18, 1912

The Boston American, April 19, 1912

The Boston Globe, Evening Edition, April 15, 1912

The Boston Globe, Morning Edition, April 16, 1912

The Boston Globe, Evening Edition, April 16, 1912

The Boston Globe, Morning Edition, April 19, 1912

The Boston Globe, Evening Edition, April 19, 1912

The Boston Globe, Morning Edition, April 21, 1912

The Cleveland Plain Dealer, Extra Edition, April 16, 1912

Daily Mirror (London) Morning Edition, April 15, 1912

Daily Mirror (London) Morning Edition, April 16, 1912

Daily Mirror (London) Extra Edition, April 16, 1912

The New York American, April 16, 1912

The New York Evening Mail, Late Edition, April 15, 1912

The New York Evening Mail, Morning Edition, April 17, 1912

The New York Evening Sun, Late Edition, April 15, 1912

The New York Times, Morning Edition, April 15, 1912

The New York Times, Evening Edition, April 15, 1912

The New York Times, Morning Edition, April 16, 1912

The New York Times, Evening Edition, April 16, 1912

The New York Times, Morning Edition, April 18, 1912

The New York Times, Evening Edition, April 18, 1912
The New York Times, Morning Edition, April 19, 1912
The New York Tribune, April 16, 1912
The New York Tribune, April 17, 1912
The New York World, April 17, 1912
Punch (Magazine) April 24, 1912
The St. Louis Post-Dispatch, April 18, 1912
Sphere (London) April 19, 1912
Sphere (London) April 20, 1912
Sphere (London) April 22, 1912
Sphere (London) April 24, 1912
Times of London, April 15, 1912
Times of London, April 16, 1912
The Virginian Pilot April 16, 1912
The Wall Street Journal, April 16, 1912

Movies & Television Shows

Too numerous to mention.
Check the Internet Movie Data Base,
when you've have an hour to kill.

Index

"One of the great defects of English books printed in the last century is the want of an index."

Lafcadio Hearn

á la carte restaurant, 141, 143, 144, 169, 206, 318, 465
A Night to Remember, 11, 500
Abbott, Rhoda, 329, 346, 447, 448, 453
Abelseth, Olaus, 185, 238
A-Deck, 155, 157, 237, 290
Admiral Beatty, 405
Admiralty, 38, 39, 41, 46, 50, 478, 487, 490
Adriatic, 120, 124, 278, 476
Aks, Leah, 285
Aks, Philip, 285
Allen, Elizabeth Walton, 297
Allison, Bessie, 286
Allison, Hudson, 286
Allison, Loraine, 286
Allison, Trevor, 286
America, 49
America's Cup, 49
American Bar Association, 447
Amerika, 33, 34, 141, 167, 204
Andersson, Anders, 354
Andrews, Elizabeth Law Barber, 117, 437
Andrews, Helen, 117, 437
Andrews, Kornelia, 305
Andrews, Thomas, 54, 55, 63, 64, 70, 116, 188, 236, 243, 260, 317, 354, 381, 392, 405, 437
   biography, 111, 112, 114, 115
   Bulkheads, 67, 70
   damage, 224, 227
   Guarantee Group, 81
   Lifeboats, 66, 67
   Olympic Maiden Voyage, 76
   Writing home, 90
Antillian, 205
Anti-Lordites, 456
Appleton, Charlotte, 350
Aquitania, 48, 49
Arcadia, 26
Archduke Franz Ferdinand, 475
Arctic, 29
Asquith, Raymond, 239
Associated Press, 354, 364
Astor IV, John J., 100, 104, 161, 164, 170, 221, 244, 300, 301, 313, 381, 390, 430, 432

Astor, Madeleine, 100, 161, 162, 163, 164, 170, 221, 244, 285, 354, 453, 469, 475
Astor, Vincent, 367
Atlantic Ferry, 26, 35, 148, 223, 475
Aubert, Léontine, 282
$B_{51}/53/55$, 259
$B_{52}/54/56$, 127, 130, 221
Baclini, Latifa, 310
Bailey, Henry Joseph, 277
Ball, Ada, 185
Ballard, Dr. Robert, 13, 466
Ballin, Albert, 141
Baltic (White Star), 120, 169, 192, 203, 204
band of brothers, 20
Barkworth, Algernon, 331
Barrett, Frederick, 291, 335
Bartholdi, Frédéric, 369
Baxter, Quigg, 100, 171, 172, 271
B-Deck, 150, 152, 157, 220, 273, 282, 287, 299, 301, 465
Becker, Marion, 287
Becker, Richard, 287
Becker, Ruth, 287, 290
Beckwith, Helen, 252
Beckwith, Richard, 252
Beckwith, Sallie, 252
Beesley, Lawrence, 290, 499
Behr, Karl, 253
Belfast, 29, 51, 55, 66, 74, 76, 81, 84, 111, 117, 137, 222, 277, 374, 438
Belfast Evening Telegraph, 137
Bell, Eng. Joseph, 317
Berthon, 421
Binns, Jack, 189
Bird, Ellen, 261
Bishop, Dick, 220
Bishop, Helen, 220
Björnström-Steffansson, Mauritz, 310, 311, 313, 350
black gang, 84, 208, 281
Blair, Lt. David, 83
Blue Ensign, 80, 106, 327, 381
Blue Riband, 31, 33, 35, 46, 49, 52, 489

Board of Trade, 11, 55, 64, 91, 92, 93, 239, 392, 404, 406, 407, 410, 411, 414, 424, 485
Boat Deck, 62, 123, 238, 247, 282, 287, 290, 299, 306, 310, 312, 318, 326
Boiler Room No. 1, 82
Boiler Room No. 4, 281
Boiler Room No. 5, 221, 227, 256
Boiler Room No. 6, 82, 221, 256
Boston, 24, 25, 134, 208, 367, 398, 408, 422, 456
Boston Red Sox, 422
Bowen, David "Dai", 97, 354
Boxhall, Lt. Joseph, 215, 228, 238, 242, 257, 288, 295, 298, 334, 339, 340, 397
Brayton, George, 283
Bride, Harold, 111, 196, 198, 199, 205, 207, 228, 254, 256, 260, 268, 302, 309, 317, 320, 323, 326, 329, 396, 397, 493
    wireless set-up, 110
Britannia, 24, 26, 381
Britannic (1874), 31
Britannic (1914), 126, 480, 492
British Registry, 328
Brown, Caroline, 350
Brown, Margaret, 100, 101, 169, 170, 258, 271, 272, 333, 477
Browne, Rev. Francis, 103
Bryan, William Jennings, 174
Buckley, Daniel, 291, 469
Bucknell College, 261
Bucknell, Emma Eliza, 261
Buley, Edward, 307
Buss, Kate, 283
Butt, Maj. Archibald, 96, 175, 176, 177, 219, 368, 430, 438
Byles, Rev. Thomas, 199, 325
Café Parisien, 145, 146, 465
Calderhead, Edward, 253
Caldwell, Albert, 141, 229, 237, 259, 290
Caldwell, Alden, 237, 290, 389
Caldwell, Sylvia, 237, 290
Caledonia, 26
Californian, 205, 208, 210, 211, 213, 218, 223, 259, 303, 397, 398, 400, 407, 409, 413, 456, 457, 465, 477, 484
Candee, Helen Churchill, 173, 174, 236, 270, 271, 350, 478, 479
Cap Arcona, 462
Cape Race, 29, 206, 207, 212, 231, 356, 357
Cardeza, Charlotte Drake, 259
Carlisle, Alexander, 55, 63, 66, 112, 392, 412
    Lifeboats, 67
Caronia, 200
Carpathia, 232, 233, 234, 235, 262, 264, 265, 266, 298, 336, 339, 340, 344, 346, 348, 349, 350, 368, 369, 383, 385, 386, 389, 394, 396, 422, 475, 480
Carter, William, 219, 310
Case, Howard, 261, 262
Cassebeer, Eleanor, 253
Cavallino Rampante, 15
C-Deck, 107, 150
Cedric, 157, 236, 386
Chambers, Bertha, 253
Chambers, Norman, 253
Chaudanson, Victorine, 301
Chelsea piers, 77, 107, 369
Cherbourg, 75, 98, 102, 164, 171, 178
Christie's, 476
Clarke, Capt. Maurice, 91, 421
Cleaver, Alice, 286
Clench, Frederick, 220, 282
Cleveland Plain Dealer, 501
Clive Cussler, 497
Cochran, Johnny, 447, 448, 453
Coffey, John, 103, 457
Collins Line, 27, 28
Collyer, Marjory, 280
Columbia, 26
Compagnie Générale Transatlantique, 32, 38
Congressional Limited, 389
Cornell, Malvina, 350
Cottam, Harold, 232, 262, 298, 350
Countess of Rothes, 95, 261, 335
Coutts, Minnie, 296
CQD, 189, 228, 232, 254, 257
Crane, Charles, 355, 364
Crawford, Alfred, 220, 332

Crosby, Captain Edward, 214
Crosby, Catherine, 249
Crosby, Harriette, 249
Cunard Line, 27, 30, 32, 38, 39, 46, 49, 57, 75, 131, 185, 364, 369
Cunard, Samuel, 24, 26, 49
Daily Mail, 42, 371, 378
Daily Mirror (London), 501
Dalbeattle, 437
Daniels, Robert, 95
Davidson, Orian, 258
Davidson, Thornton, 258
Davies, Elizabeth, 278
Davies, Joseph, 278
D-Deck, 138, 146, 154, 205, 222
Del Carlo. Argene, 286
DeMessemaeker, Anna, 290
Deutschland, 33, 34
Dick, Vera, 259
Dodge, Dr. Washington, 185, 253, 291
Dodge, Ruth, 253
Dollar Princesses, 426
Douglas, Donald, 297
Douglas, Mahala, 297
drydock, 52, 55, 81, 457
Duff-Gordon, Lady Lucy, 100, 205, 267, 337
Duff-Gordon, Sir Cosmo, 100, 177, 178, 266, 267, 337, 409
Dyamond, Frank, 292
Dyker, Adolph, 274
Dyker, Anna, 274
Earl of Balfour, 39, 52
E-Deck, 154, 221
Ellis Island, 369
Englehardt, 64, 296, 309, 313, 349
Evans, Cyril, 208, 212, 213, 223, 260, 400, 402
Fastnet, 106, 193, 480
Faulkner, William, 286
F-Deck, 148, 149, 152, 154, 221
Fenway Park, 422
Fin de siècle, 425
First Class passengers, 107, 143, 146, 150, 161, 208, 310, 381
    Dining Room, 141
    gymnasium, 134
    squash court, 136
    swimming pool, 137
    turkish bath, 136
Fleet Street, 369, 404, 493
Fleet, Frederick, 216, 272
    crow's nest, 214
    Iceberg, 217
Fletcher, Peter W., 147
Florida, 189
Fortune, Mark, 306
Francatelli, Laura, 178, 267
France, 203
Franklin, P. A. S., 201, 360
Frauenthal, Dr. Henry, 253
Frick, Henry Clay, 127
Frölicher, Magaritha, 237
Futility, 110
Futrelle, Jacques, 257, 283, 312, 313
Futrelle, Lily, 283, 312
Gallipoli, 480, 489
Gardiner, Robin, 464
Gatti, Gaspare Antonio Pietro, 147, 354
G-Deck, 154
Gibson, Dorothy, 220, 249, 250, 251, 332
Gibson, Henry, 303, 305
Gibson, James, 260
Gigantic, 53, 127, 225, 452, 480, 489
Giglio, Victor, 283
Gill, Ernest, 398, 407
Gimbels, 253
Ginger Rogers, 415
Gloucester, 214
Goebbels, Joseph, 461
Gorta Mór, 104
Gospel of Mammon, 19
Götterdämmerung, 326
Gracie, Col. Archibald, 96, 173, 214, 224, 269, 270, 271, 300, 332, 341, 350
Graham, Edith, 261, 262
Graham, Margaret, 261
Grand Banks, 214
Great Coal Strike, 82, 292, 378
Groves, Lt. Charles, 223, 259, 260, 400, 408
growlers, 14, 215, 339
Guggenheim, Benjamin, 100, 164, 165, 282, 284, 354
Guggenheim, Florette, 164, 367

Haddock, Capt. Herbert, 124, 421
Haines, Albert, 229, 284
Hakkarainen, Elin, 292
Halifax, 127, 196, 356, 357, 364, 367, 426, 480
Halifax Morning Chronicle, 127
Hamburg-Amerika Line, 33, 34
Harding, J. Horace, 127
Hardy, Thomas, 442
Harland & Wolff, 29, 51, 55, 57, 58, 61, 66, 74, 75, 81, 93, 111, 114, 185, 222, 252, 437, 447, 464, 465, 481
Harper, Henry Sleeper, 100, 259
Harper, Myra, 259
Hart, Benjamin, 278
Hart, Esther, 278
Hart, Eva, 278, 481
Hartley, Wallace, 139, 182, 183, 238, 293, 323, 354, 430, 431
Harvard, 433, 463, 464
Harvey, Eng. Herbert, 256
Hays, Charles, 214, 258
Hays, Clara, 258
Hays, Margaret, 249
Hearst, William Randolph, 422, 423, 424, 482
Hecht, Ben, 424, 425, 482
Hemming, Samuel, 229
Herkner, Anna, 156
Hess, Mrs. Alfred, 367
Hichens, Robert, 272, 333, 405, 407, 482
    Quartermaster, 215, 217
HMS Dreadnought, 46
HMS Hawke, 81, 123, 181, 464
HMS Victory, 20, 492
Hogeboom, Anna, 305
Hogg, George, 251
Homeric, 492
Horatio Alger, 385
Hutchinson, Jim, 206, 224
hypothermia, 329, 346
Iceberg, 29, 217, 220, 222, 224, 227, 228, 285, 349, 368, 369, 392, 402, 413, 444, 445, 452, 466
Iceberg Alley, 193, 214
International Mercantile Marine, 36, 37, 38, 52, 84, 168, 364, 365, 412, 447, 451

Irish Masters, 56, 393
Isle of Wight, 49, 374
Ismay, J. Bruce, 29, 30, 50, 51, 52, 63, 67, 70, 97, 128, 129, 152, 169, 203, 204, 221, 224, 227, 252, 253, 254, 310, 324, 346, 380, 389, 390, 391, 392, 393, 395, 400, 403, 409, 410, 411, 412, 422, 424, 461, 483
    biography, 128, 131
    Launch, 75
    Olympic, 77
    Speed record, 201
Ismay, Thomas, 29, 30, 31, 51, 131, 493
Jarrow, 489, 490
Jensen, Carla, 274
Jerwan, Marie, 203, 287
Jessop, Violet, 181, 276, 480
Jewel, Archie, 251
Jimmy Cagney, 10
Johnson, James, 237
Joslyn, Allyn, 470
Joughin, Charles, 332, 341, 405
Kaiser Wilhelm der Grosse, 33
Kaiser Wilhelm II, 34, 35
Kate Winslet, 12
Kent, Edward, 270, 313, 354
Kimball, Edwin, 253
Kimball, Gertrude, 253
King Edward VII, 425, 427
Kipling, Rudyard, 47, 134
Kreuchen, Emilie, 297
La Belle Epoque, 425
La Cuesta Encantada, 482
La Touraine, 196
Lee, Reginald
    crow's nest, 214
Lévy, René Jacques, 202, 203, 287
Leyland Line, 208
Liberty Enlightening the World, 369
lifebelt, 242, 244, 292, 301, 309
Lifeboat A (collapsible), 313, 318, 332, 346
Lifeboat B (collapsible), 313, 326, 329, 330, 332, 340
Lifeboat C (collapsible), 309, 310, 311, 312, 313
Lifeboat D (collapsible), 311, 313, 341, 346
Lifeboat No. 1 (cutter), 266, 267

Lifeboat No. 10, 299, 305, 307
Lifeboat No. 11, 285, 286, 287, 288, 292
Lifeboat No. 12, 282, 340, 349
Lifeboat No. 13, 290, 291, 292, 307, 335
Lifeboat No. 14, 277, 280, 341, 346
Lifeboat No. 15, 292, 293
Lifeboat No. 16, 273, 276
Lifeboat No. 2 (cutter), 296, 297, 298, 334, 340
Lifeboat No. 3, 258, 259, 266, 346, 409
Lifeboat No. 4, 299, 300, 301, 302, 305, 310, 340
Lifeboat No. 5, 252, 253, 254, 267, 291, 327, 333, 409
Lifeboat No. 6, 270, 271, 272, 273, 277, 299, 333, 405
Lifeboat No. 7, 248, 252
Lifeboat No. 8, 260, 262, 271, 299, 335
Lifeboat No. 9, 262, 266, 270, 271, 273, 282, 285
Lightoller, Lt. Charles, 83, 92, 93, 206, 207, 228, 229, 231, 251, 257, 260, 261, 262, 268, 270, 272, 273, 277, 281, 296, 297, 298, 299, 300, 301, 311, 314, 318, 330, 332, 340, 349, 394, 395, 397, 403, 441, 484, 500
Liverpool, 24, 27, 29, 75, 120, 374, 405, 436, 437, 438, 480, 487
Lloyd, Edward, 327
Lloyd's of London, 33, 327, 366
locus standi, 451
Longley, Gretchen, 305
Lord Mersey, 404, 406, 413
Lord Nelson, 21, 49, 106, 336
Lord Pirrie, 51, 55, 122
  Lifeboats, 67
Lord, Capt. Stanley, 208, 223, 260, 305, 400, 402, 403, 413, 414, 485
Lord, Walter, 11, 231
Lowe, Lt. Harold, 251, 252, 254, 277, 278, 310, 334, 338, 347, 485
Lusitania, 30, 39, 42, 44, 45, 46, 49, 57, 142, 487, 489
Lutine Bell, 327
Mackay-Bennett, 323, 426, 428, 430, 479
Macy's, 95, 167, 434
Madill, Georgette Alexandra, 297
mailroom, 105

Majestic (1890), 31
Majestic (1922), 492
mal de mer, 134, 237
Malachard, Jean-Noël, 203
Marconi, Guglielmo, 110, 189, 191, 231, 400
Maréchal, Pierre, 251
Marsden, Evelyn, 276
Marshall, Charles H., 350
Mauretania, 30, 39, 43, 44, 46, 47, 52, 57, 58, 127, 200, 459, 488, 491
Maxim, 425
Mayné, Bertha, 100, 171, 271
McCawley, Thomas W., 134
McCoy, Agnes, 274
McCoy, Alice, 274
McCoy, Bernard, 274
McElroy, Hugh, 90, 134, 309, 354
McGhee, Dr. Frank, 346
McGough, George, 284
McGough, James, 219, 251, 285
Mewes, Charles, 141
Millet, Francis Davis, 105
Moody, Lt. James, 215, 228, 273, 274, 276, 289, 291, 318, 354
Moor, Beila, 278
Moore, Clarence, 219
Moore, George, 259
Morgan, J. P., 19, 36, 37, 50, 108, 127, 131
Morrison, Stephan G., 448
Mount Temple, 351
Mulvihill, Bertha E., 292
Murdoch, Lt. William, 12, 13, 83, 197, 198, 207, 215, 217, 228, 248, 251, 252, 254, 259, 262, 266, 267, 282, 283, 285, 288, 289, 290, 291, 305, 307, 309, 314, 318, 354, 437, 466, 493
  Iceberg, 217, 222
Murphy, Kate, 276
Murphy, Maggie, 276
Nantucket, 189, 196, 231
Navratil, Michel, 179, 180, 181, 312
Nearer My God To Thee, 11, 185, 323, 416
Neuengamme, 462

New York, 24, 97, 106, 120, 123, 134, 164, 196, 232, 354, 360, 364, 389, 439
New York Stock Exchange, 364
New York Times, 383
Newlands, Francis, 390
Newport, Rhode Island, 20
Newsom, Helen, 253
Nicola-Yarred, Elias, 310
Nicola-Yarred, Jamila, 310
No. 18 Broad Street, 364
No. 219 Madison Avenue, 36
No. 24 Belgrave Square, 50, 483
No. 27 Chelsea Street, 50
No. 9 Broadway, 360, 365, 374
Nomadic, 98, 481
Nomadic,, 57
Noordam, 199
Norman, Douglas, 283
Normandie, 459
Norris, Richard, 332
North German Lloyd Line, 33, 35
O God, Our Help in Ages Past, 323
O'Loughlin, Dr. William, 268
Occam's razor., 457
Oceanic, 29, 31, 97, 131, 476, 490, 492
Oceanic House, 373
Old Head of Kinsale, 106, 487
Olympic, 90, 181, 254, 276, 303, 350, 357, 421, 452, 464, 465, 490
   accommodations, 150
   Design, 53, 56, 57, 62, 66
   HMS Hawke collision, 81
   Launch, 74
   Maiden Voyage, 76
   Publicity, 64
   Restaurant, 142
Olympic-class, 14, 53, 61, 68, 84, 111, 124, 201, 313, 480
   Design, 63
   Shipbuilder, 65
Orlop Deck, 229
our coterie, 174, 270, 310
Pacific, 29
Pain, Dr. Alfred, 283
Parable of the Unjust Steward, 207
Paris, 92, 164, 203, 477, 491
Parsons, Charles, 38, 41, 46, 47, 57, 319, 491

Patterson, James, 456
Pax Britannica, 18, 33, 426
Pax Germanica, 33
Perkis, Walter, 302
Peter-Joseph, Catherine, 310
Peter-Joseph, Michael, 310, 311
Peuchen, Maj. Arthur Godfrey, 95, 185, 237, 270, 272, 333
Philadelphia, 19, 95, 104, 164, 167, 299, 451, 458
Phillips, John, 196, 198, 199, 204, 206, 207, 208, 212, 228, 229, 232, 238, 247, 254, 255, 260, 268, 285, 302, 303, 309, 317, 318, 319, 320
   Californian, 212
   SOS, 255
   wireless set-up, 110
Pier 43, 84
Pier 44, 84, 95, 96, 421
Pier 54, 45, 369, 370, 389
Pier 59, 107, 369
Pitman, Lt. Herbert, 228, 230, 231, 252, 254, 327, 333, 490
Plymouth, 24
Poigndestre, John, 282
Prince Albert, 18, 19
Prinz Adalbert, 349
propellers, 58, 81, 123, 222, 349, 417
Punch, 381, 382, 502
Pusey, Robert, 337
Quarantine, 369
Queen Victoria, 18, 49
Queen's Island, 51, 56, 61, 74, 222
Queenstown, 101, 103, 105, 106, 158, 237, 438, 487
Raise the Titanic!, 497
Rappahannock, 196, 198
Ray, F. Dent, 290
Ray, Frederick, 290
Red Ensign, 106
Republic, 189
Ritz, Cesar, 141
Ritz-Carlton, 141
River Clyde, 46, 49, 489
River Lagan, 76
River Marne, 425
River Mersey, 483
River Tyne, 39, 489
RMS Queen Elizabeth, 475

RMS Queen Mary, 458, 489
Robert, Elizabeth, 297
Robertson, Morgan, 110, 441
Roche's Point, 101
Roebling, Augustus, 261
Romaine, Charles, 283
Roman Empire, 426
Roosevelt, Theodore, 96, 176, 355
Rosenbaum, Edith Louise, 287
Rosenshine, George, 100, 312
Rostron, Capt. Arthur, 336, 350, 491
   biography, 236
   lifeboats, 339, 340, 349
   more power, 298
   orders, 262, 264, 265, 266
   U.S. Senate, 393, 394
Rothschild, Elizabeth, 271
Rowe, George, 257, 262, 271, 295, 311
Royal Navy, 18, 20, 33, 36, 41, 66, 106, 464, 481, 487
Rubáiyát of Omar Khayyám, 101
Rudyard Kipling, 394
Ryan, Edward, 469
Ryerson, Arthur, 301, 304
Ryerson, Emily, 203, 301, 302
Ryerson, Jack, 301
Sägasser, Emma, 282
Salomon, Abraham Lincoln, 267
San Francisco Earthquake, 220
Sanderson, Harold, 412
Scarrott, Joseph, 280
Schwabe, Gustav, 29
Scotland Road, 221, 238
Second Class passengers, 32, 75, 95, 101, 106, 147, 181, 199, 273, 290
Seward, Frederick K., 251
Shakespeare, William, 317, 353, 449, 454, 474, 482
Shepherd, Eng. Jonathon, 256
Shipbuilder, 65, 66, 71, 116, 150, 445, 499
Siasconset, 190, 231
Sir Walter Raleigh, 19
Sloper, William T., 251
Smith, Capt. Edward J., 81, 83, 95, 98, 106, 117, 186, 203, 251, 262, 272, 277, 311, 320, 326, 354, 403, 413, 438, 493
   biography, 120
   damage, 225
   speed, 200
Smith, William Alden, 385, 389, 390, 392, 393, 394, 400, 403, 404, 409
Social Register, 383
Solent, 374
Southampton, 75, 81, 82, 84, 95, 101, 124, 167, 179, 249, 276, 278, 305, 374, 378, 379, 380, 421, 437, 438, 465
Spedden, Frederic Oakley, 259
Speeden, Daisy, 259
Sphere (London), 502
SS Frankfurt, 247, 303
Statue of Liberty, 369
Stengel, Annie, 253
Stengel, Charles Henry, 253, 267
Stephenson, Martha, 220
Stone, Lt. Herbert, 259, 260, 303
Stone, Martha Evelyn, 272
Straus, Ida, 95, 167, 261, 367, 433
Straus, Isidor, 95, 166, 167, 261, 367, 433
Sun Yat Sen, 259
Symons, George, 267
Taft, William Howard, 96, 176, 177, 355, 389, 430
Telefunken, 302
Teutonic, 33
Thayer, Jack, 100, 241, 243, 321, 329, 341, 421, 500
Thayer, John, 100, 243
Thayer, Marian, 100, 203, 300, 301
The Belfast News, 76
The Big Broadcast of 1938, 481
The Boston American, 398, 500
The Boston Globe, 422, 500, 501
The Boston Tea Party, 27
The Chicago Daily News, 482
The Gilded Age, 19, 141
The Great Circle Route, 192, 193
The Irish News and Belfast Morning News, 66
The Marconi Company, 12, 111, 189, 207, 302, 364
The Mikado, 414
The New York American, 501
The New York Evening Mail, 501
The New York Evening Sun, 501

The New York Times, 124, 258, 355, 356, 358, 383, 422, 432, 434, 441, 501, 502
The New York Tribune, 502
The New York World, 502
The Plaza Hotel, 272
The Truth about Chickamauga, 13, 174
The Unsinkable Molly Brown, 477
Third Class accommodations, 154
Third Class Dining Rooms, 148
Third Class passengers, 32, 61, 95, 101, 106, 152, 221, 273, 285, 309, 317, 441
Thomas Andrews, Shipbuilder, 116
Thomas, Charles, 277
Thompson Graving Dock, 55, 73, 78, 79, 81
Thorne, Maybelle, 100, 312
Times of London, 41, 502
Tipperary, 158
Titan, 110
Titan-Carpath, 383
Titanic, 276, 336, 351, 357, 369, 383, 398, 408, 413, 430, 452, 456, 459, 461, 464, 484, 490
   accommodations, 139, 151, 152, 154, 156, 157
   Coal, 82, 84
   CQD, 231
   damage, 221, 225, 227, 466, 467
   Design, 53, 55, 62, 64, 66
   Dining, 139
   engine room flooding, 298
   Engineer's Memorial, 436
   Fitting out, 81
   Food, 87
   German movie, 461
   Ice warning, 196, 199, 203, 204, 206
   Iceberg, 217
   Launch, 75
   Leaving Cherbourg, 101
   Leaving Queenstown, 106
   Leaving Southampton, 97
   lifeboats, 252
   Publicity, 64
   Restaurant, 142
   service, 152
   sinking, 318

SOS, 255
speed, 200
weather, 134
wireless, 110, 192, 198
Titanic (1953), 11, 470
Titanic (1997), 12, 13
Titanic (Germany, 1943), 462
Titanic Historical Society, 5, 11, 467, 481
Toronto, 447
Trachoma, 95
Trafalgar, 21, 336
Trafalgar Square, 36
Traffic, 98
Trout, Winnie, 276
Turbinia, 13, 40, 41, 491
Turja, Anna, 293
Twain, Mark, 19
U. S. Navy, 385, 490, 497
U. S. Senate, 11, 161, 170, 251, 385, 394, 397, 398, 399, 400, 401, 402, 403, 409
U-103, 490
U-20, 487
U-35, 478
U-55, 480
UCLA, 466
United States, 459, 490
United States Post Office, 105
USS Chester, 368
Van Anda, Carr, 355, 356, 357, 360, 441, 492
Vanderbilt, Alfred Gwynne, 131, 132
Vinolia Soap, 72
Virginian, 318
Waldorf-Astoria, 161, 390, 391
Wall Street Journal, 440, 502
Wanamaker's, 12, 356
War of 1812, 27
Waterloo Station, 36, 95
Watson, Ennis, 185
Welin, 63, 244, 247, 254
Wells, H.G., 355
West, Ada, 306
White Ensign, 106
White Star Line, 30, 32, 33, 38, 49, 51, 57, 74, 75, 120, 125, 128, 136, 138, 147, 148, 189, 339, 360, 374, 389, 414, 430, 447, 448, 491, 493

Boarding *Titanic*, 95
Coal, 82
Lifeboats, 63
Publicity, 64
Widener, Eleanor, 95, 167, 301, 449, 451, 463
Widener, George, 95, 167, 463
Widener, Harry Elkins, 95, 168, 219, 310, 463
Widener, Harry Elkins Memorial Library, 463
Wilde, Lt. Henry, 12, 54, 83, 84, 105, 227, 260, 277, 281, 297, 299, 309, 311, 354, 493

Wilding, Edward, 251
Wilhelm II, Kaiser, 33, 35, 41
William of Occam., 457, 458
Witter, James, 287
Woolner, Hugh, 247, 270, 271, 310, 311, 313, 350
Wright, Marion, 283
Yankee Doodle Dandy, 10
Yellow Cab Company, 454
Young, Marie, 261
zeitgeist, 425

www.ingramcontent.com/pod-product-compliance
Lightning Source LLC
Chambersburg PA
CBHW081341180526
45171CB00006B/575